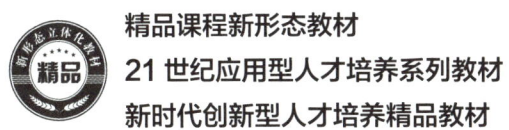

精品课程新形态教材
21世纪应用型人才培养系列教材
新时代创新型人才培养精品教材

SSM框架教程

主编 张 超 贺霖龄 张 晶

湖南大学出版社·长沙

内 容 简 介

本书共 15 章,第 1 章是环境介绍,主要讲解 JDK、IDEA、Maven 等的安装。第 2~5 章主要讲解了 MyBatis 的相关知识,其中包含初识 MyBatis、MyBatis 的核心配置、动态 SQL、MyBatis 的关联关系。第 6~9 章主要讲解 Spring 的基本应用、Spring 中的 Bean、Spring AOP、MyBatis 与 Spring 的整合。第 10~14 章主要讲解了 Spring MVC 的相关知识,其中包含 Spring MVC 入门、Spring MVC 的核心类和注解、数据绑定、SSM 框架整合。第 15 章是整个 SSM 框架的总结与综合运用,全章通过一个教务管理系统案例,贴近实际地讲解了开发中 SSM 框架的应用。

本书既可作为计算机相关专业的程序设计教材,也可作为 Java 技术的培训图书,还适合广大编程爱好者阅读与使用。

图书在版编目(CIP)数据

SSM 框架教程/张超,贺霖龄,张晶主编. —长沙：
湖南大学出版社, 2023.1(2024.8 重印)
ISBN 978-7-5667-2750-3

Ⅰ.①S… Ⅱ.①张… ②贺… ③张… Ⅲ.①JAVA 语言-程序设计
Ⅳ.①TP312.8

中国版本图书馆 CIP 数据核字(2022)第 222309 号

SSM框架教程
SSM KUANGJIA JIAOCHENG

主　　编：	张　超　贺霖龄　张　晶
责任编辑：	黄　旺
印　　装：	涿州汇美亿浓印刷有限公司
开　　本：	787 mm×1092 mm　1/16　印　张：15　字　数：316 千字
版　　次：	2023 年 1 月第 1 版　印　次：2024 年 8 月第 2 次印刷
书　　号：	ISBN 978-7-5667-2750-3
定　　价：	45.00 元

出 版 人：李文邦
出版发行：湖南大学出版社
社　　址：湖南·长沙·岳麓山　邮　编：410082
电　　话：0731-88822559(营销部),88820006(编辑室),88821006(出版部)
传　　真：0731-88822264(总编室)
网　　址：http://press.hnu.edu.cn
电子邮箱：1176142336@qq.com

版权所有,盗版必究
图书凡有印装差错,请与营销部联系

《SSM 框架教程》编委会

主　编：张　超　贺霖龄　张　晶
副主编：陈文驰　曾回生　周　华　陈春晖　熊志文

Foreword 前言

党的二十大报告指出，加快发展数字经济，促进数字经济和实体经济深度融合。新一代信息技术与各产业结合形成数字化生产力和数字经济，是现代化经济体系发展的重要方向。

本书详细讲解了 JavaEE 中 Spring、Spring MVC 和 MyBatis 三大框架（以下简称"SSM"）的基本知识和应用。本书在对知识点进行描述时采用了大量案例，可以更好地帮助读者学习和理解 SSM 的核心技术。

本书详细介绍了 IDEA 的使用，并为每章精心设计了思政元素，因势利导，依据专业课程的特点，采取了恰当方式，自然融入中华传统文化和职业素养等元素。注重挖掘课程中的思政教育要素，弘扬职业精神和工匠精神。

本书共 15 章，第 1 章是环境介绍，主要讲解 JDK、IDEA、Maven 等的安装。第 2~5 章主要讲解了 MyBatis 的相关知识，其中包含初识 MyBatis、MyBatis 的核心配置、动态 SQL、MyBatis 的关联关系。第 6~9 章主要讲解 Spring 的基本应用、Spring 中的 Bean、Spring AOP、MyBatis 与 Spring 的整合。第 10~14 章主要讲解了 SpringMVC 的相关知识，其中包含 Spring MVC 入门、Spring MVC 的核心类和注解、数据绑定、SSM 框架整合。第 15 章讲解整个 SSM 框架的总结与综合运用，全章通过一个教务管理系统案例，贴近实际地讲解了开发中 SSM 框架的应用。读者掌握了 SSM 框架技术，就能很好地适应企业开发的技术需求，为大型项目的开发奠定基础。

本书既可作为计算机相关专业的程序设计教材，也可作为 Java 技术的培训图书，适合广大编程爱好者阅读与使用。

由于编者水平有限，书中难免存在疏漏和不足之处，敬请读者批评指正。

编 者

Contents 目录

第 1 章 环境介绍 1
1.1 环境介绍 1
1.2 安装 JDK 2
1.3 安装 IDEA 6
1.4 安装 Maven 9

第 2 章 初识 MyBatis 14
2.1 简介 14
2.2 特点 15
2.3 MyBatis 的工作原理 15
2.4 入门案例 17

第 3 章 MyBatis 的核心配置 24
3.1 配置文件的主要元素介绍 25
3.2 核心类介绍 26
3.3 映射文件的主要标签介绍 28

第 4 章 动态 SQL 31
4.1 动态 SQL 介绍 32
4.2 动态 SQL 标签介绍 32
4.3 动态 SQL 标签案例 32

第 5 章 MyBatis 的关联关系 37
5.1 关联关系概述 37
5.2 一对一 39
5.3 一对多 46

第 6 章 Spring 的基本应用 53
6.1 认识 Spring 框架 53
6.2 Spring 框架优点 54
6.3 Spring 框架体系结构 54
6.4 Spring 项目布局 57
6.5 Spring 核心容器 61
6.6 Spring 入门程序 63

第 7 章 Spring 框架之 IoC ... 65

7.1 Spring 操作 Bean ... 65
7.2 创建 Bean ... 66
7.3 管理 Bean ... 68
7.4 依赖注入 ... 69

第 8 章 Spring 之 AOP ... 78

8.1 AOP ... 78
8.2 AOP 术语 ... 79
8.3 AOP 配置及实现 ... 79

第 9 章 MyBatis 与 Spring 的整合 ... 88

9.1 整合 jar 包介绍 ... 88
9.2 案例 ... 93
9.3 利用 Springtest 模块进行测试 ... 97

第 10 章 Spring MVC 框架 ... 100

10.1 Spring MVC 简介 ... 100
10.2 Spring MVC 的优点 ... 101
10.3 Spring MVC 框架工作原理 ... 101
10.4 Spring MVC 核心类 ... 103
10.5 第一个 Spring MVC 程序 ... 105

第 11 章 Spring MVC 框架注解和数据传递 ... 113

11.1 Spring MVC 框架常用注解 ... 113
11.2 Controller 注解类型 ... 114
11.3 RequestMapping 注解类型 ... 114
11.4 @RequestParam 注解 ... 115
11.5 请求处理方法的参数类型和返回类型 ... 115
11.6 视图解析器 ViewReSolver ... 117
11.7 数据绑定 ... 117

第 12 章 JSON 数据交互和 RESTful 支持 ... 121

12.1 JSON 概述 ... 121
12.2 JSON 数据转换 ... 122
12.3 RESTful 支持 ... 128
12.4 应用案例——用户信息查询 ... 129

第 13 章 拦截器 ... 132

13.1 拦截器概述 ... 132
13.2 拦截器的执行流程 ... 134

13.3　应用案例——用户认证拦截 …………………………………………… 137

第14章　SSM框架整合 ………………………………………………………… 143
14.1　SSM整合jar包介绍 …………………………………………………… 144
14.2　整合步骤 ……………………………………………………………… 147

第15章　教务管理系统 …………………………………………………………… 155
15.1　系统概述 ……………………………………………………………… 155
15.2　系统架构设计 ………………………………………………………… 156
15.3　文件组织结构 ………………………………………………………… 157
15.4　系统开发及运行环境 ………………………………………………… 158
15.5　数据库设计 …………………………………………………………… 158
15.6　系统环境搭建 ………………………………………………………… 160
15.7　功能实现 ……………………………………………………………… 171
15.8　创建页面视图（view） ……………………………………………… 207
15.9　创建控制层（controller） …………………………………………… 212

参考文献 …………………………………………………………………………… 229

第1章 环境介绍

 学习目标

[本章知识点]
  安装 JDK
  安装 IDEA
  安装 Maven

[思政目标]
  使学生理解并重视工匠精神,在学习中努力发扬工匠精神。
  使学生了解计算机软件从业人员应当具备的职业道德守则,为进军软件行业做准备。

1.1 环境介绍

为了确保在使用本书时不出现版本问题,请读者在学习之前检查开发设备环境,与表1-1所示版本对照。如有未安装版本,请及时安装。

表1-1 开发环境

环境名	版本号
JDK	1.8
IDEA	2021-2
Maven	3.6.3
MySQL	5.0

1.2 安装 JDK

去官网下载 JDK 安装包,如图 1-1 所示。

图 1-1　JDK 安装包

下载完后,双击进行安装,如图 1-2 所示。

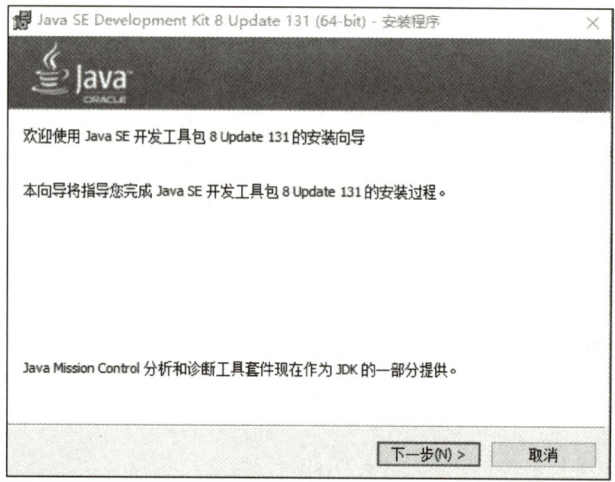

图 1-2　JDK 安装

安装过程很简单,只需要改一下安装位置(也可以不改,但要确定记得这个位置,环境变量配置时要用),然后全部点击下一步即可,如图 1-3 所示。

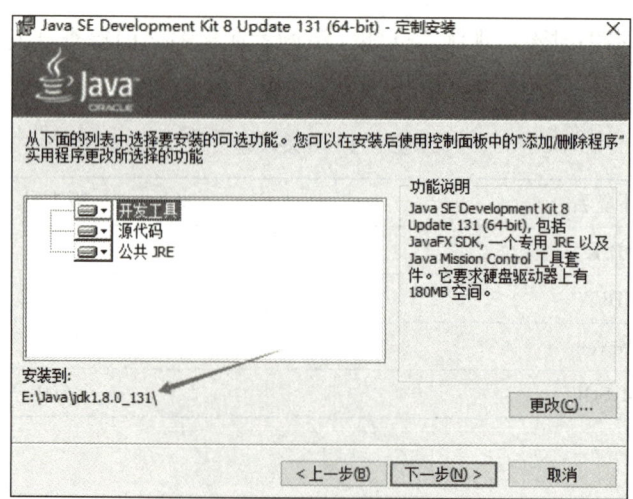

图 1-3　JDK 定制安装

设置安装目录如图 1-4 所示。

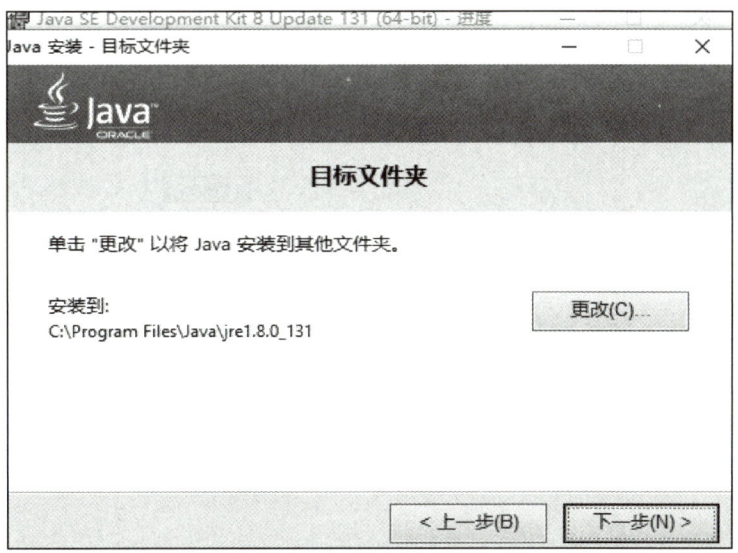

图 1-4　路径选择

JDK 安装成功画面如图 1-5 所示。

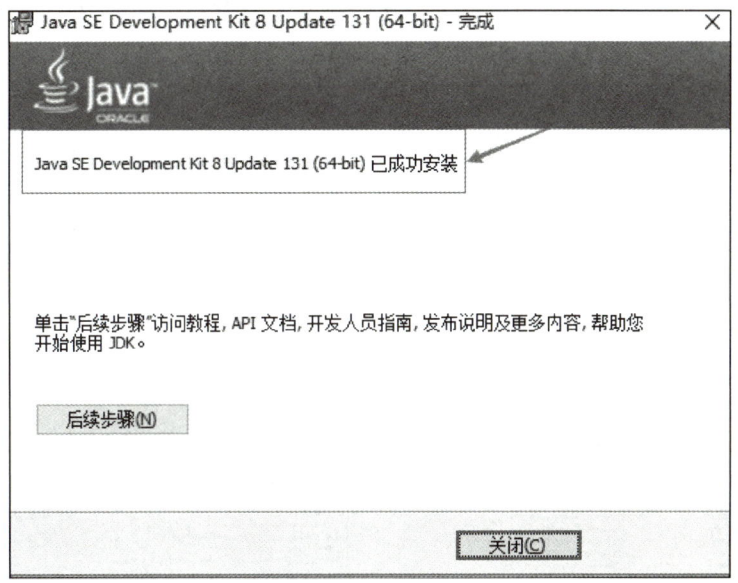

图 1-5　安装完成

安装完成后，需要进行环境变量的配置，右键点击我的电脑—属性—高级，如图 1-6 所示。

点击图 1-6 中的环境变量，然后开始环境变量的配置。

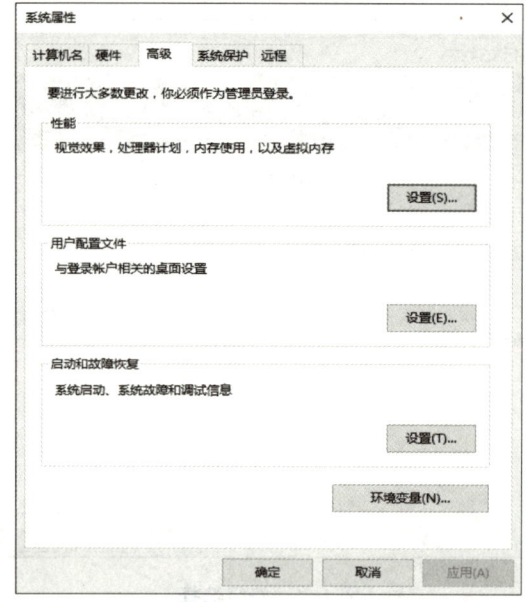

图 1-6　环境变量

点击系统变量下面的新建按钮，变量名输入"JAVA_HOME"，值对应的是你的 JDK 的安装路径，如图 1-7 所示。

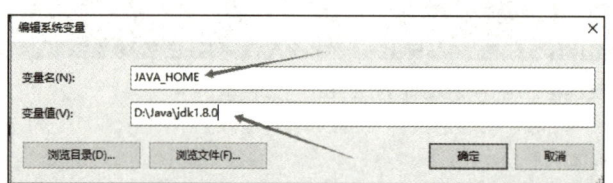

图 1-7　JAVA_HOME 设置

继续在系统变量里面双击 Path 变量，将出现如图 1-8 右边所示面板，单击新建，输入"%JAVA_HOME% \ bin"。

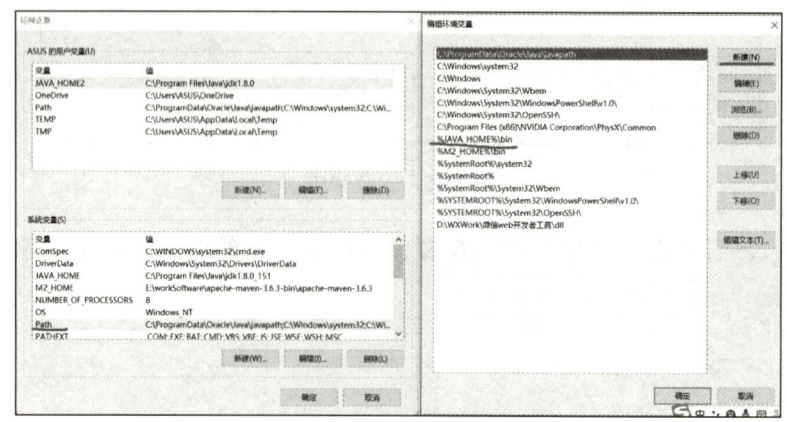

图 1-8　配置环境变量

测试自己所配置的环境变量是否正确，打开如图 1-9 所示命令行。

图 1-9　测试环境变量

输入 java -version 命令，会出现如图 1-10 所示的提示，你可以查看安装的 JDK 版本。

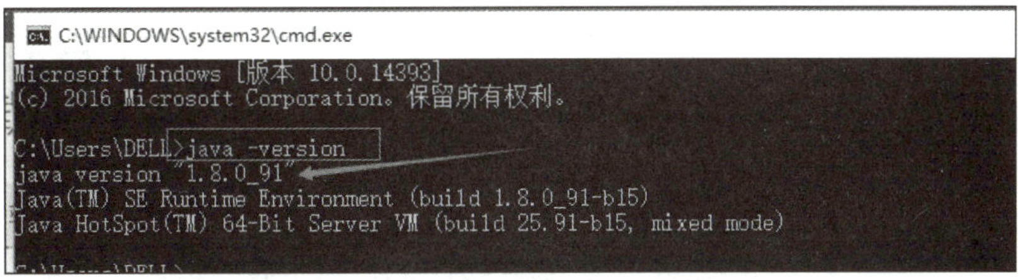

图 1-10　查看 JDK 安装版本

输入 javac 命令会出现如图 1-11 所示的提示。

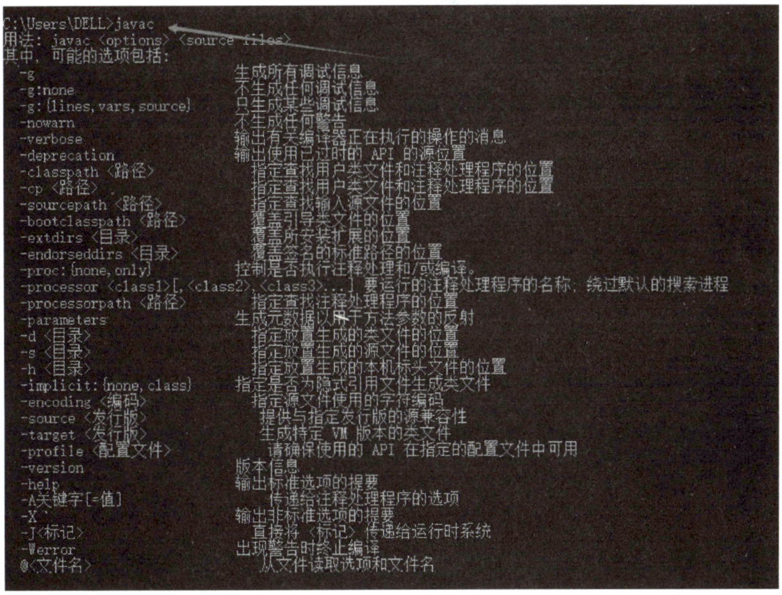

图 1-11　javac 命令

上面两项命令显示正常则表示环境变量配置成功。

1.3 安装 IDEA

1.3.1 安装 IDEA

IDEA 是学习 Java 的一种编译器,这种编译器需在官网上下载,下载地址:jetbrains.com/idea/。如图 1-12 所示。

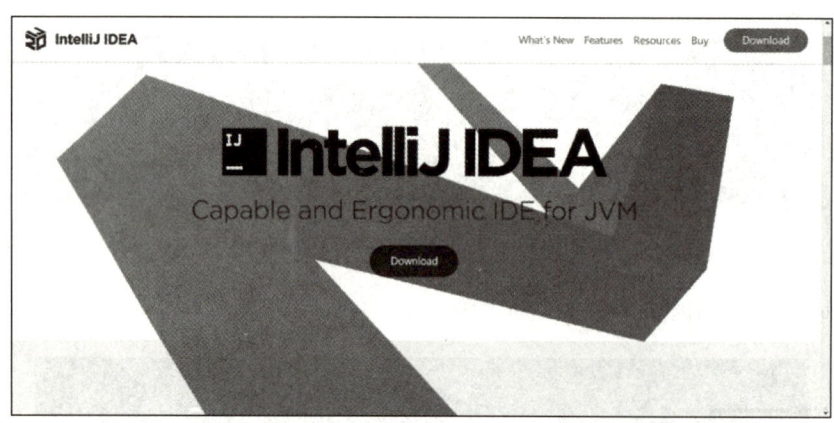

图 1-12 IDEA 下载

IDEA 有两种版本:一种是旗舰版,功能齐全,但是只有 30 天试用期;还有一种是社区版,功能不齐全,但是一般足够使用。大家可以自己选择,如图 1-13 所示。

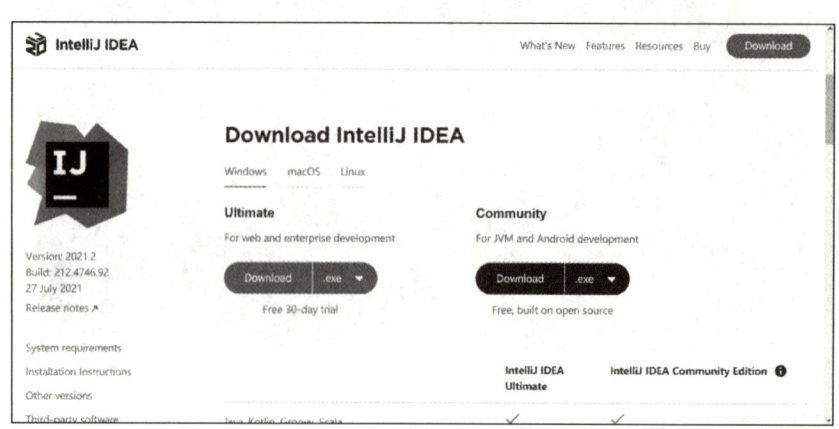

图 1-13 版本选择

第 1 章 环境介绍

1.3.2 安装教程

双击下载的 IDEA，然后点击 Next，如图 1-14 所示。

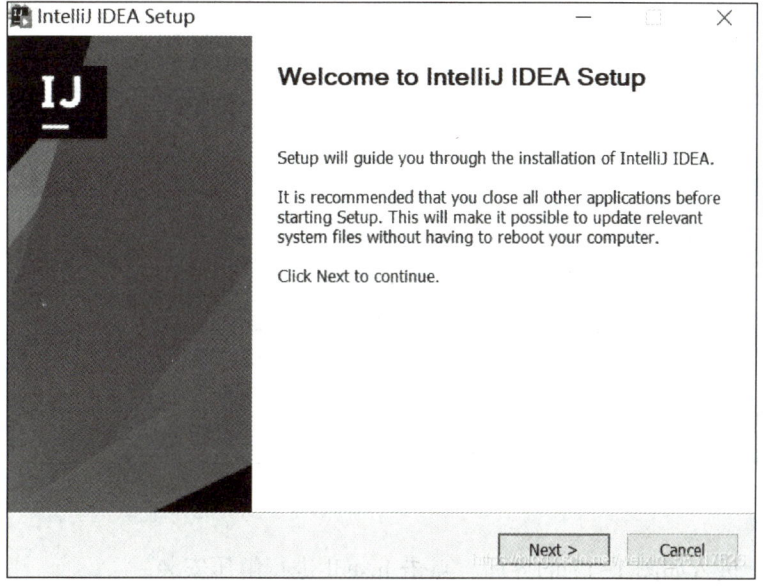

图 1-14 IDEA 安装

选择安装路径，如图 1-15 所示。

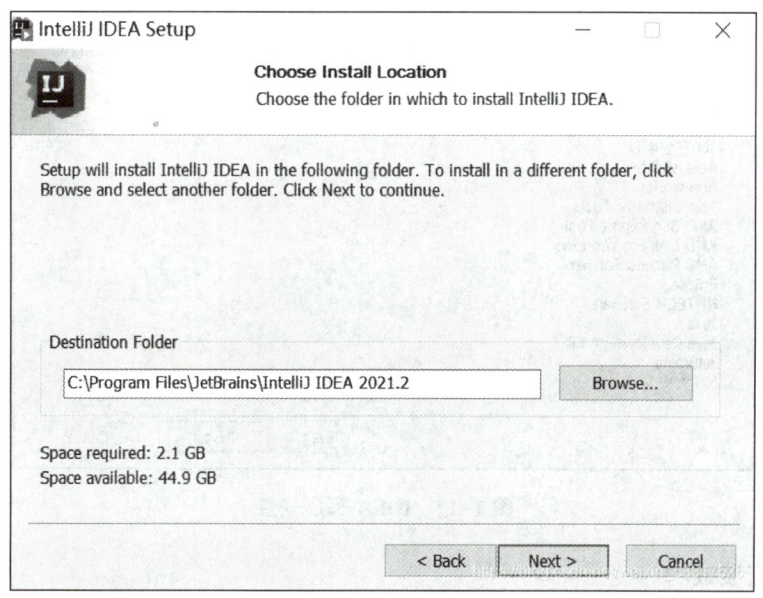

图 1-15 路径选择

然后创建快捷方式，根据自己的情况勾选，但是第一行的第一个必须勾选，这表示创建

— 7 —

桌面快捷方式，如图 1-16 所示。

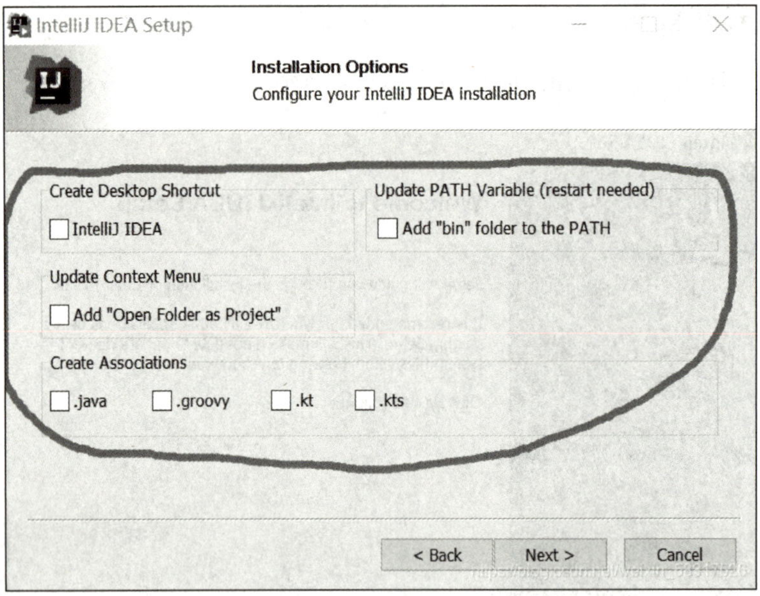

图 1-16　创建快捷方式

点击 Next，进入如图 1-17 的界面，点击 Install 进行解压安装。

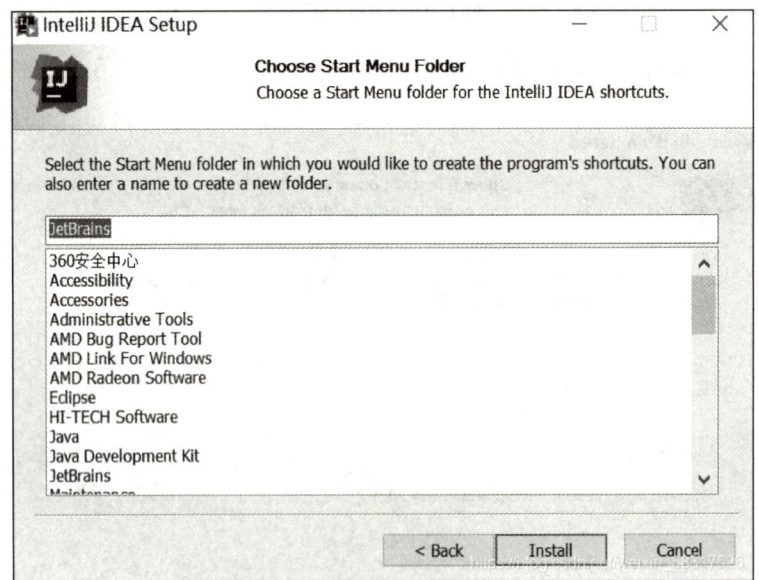

图 1-17　IDEA 解压安装

然后等待安装成功。

1.4 安装 Maven

1.4.1 安装 Maven

前往 https://maven.apache.org/download.cgi 下载需要的 Maven 版本，如图 1-18 所示。

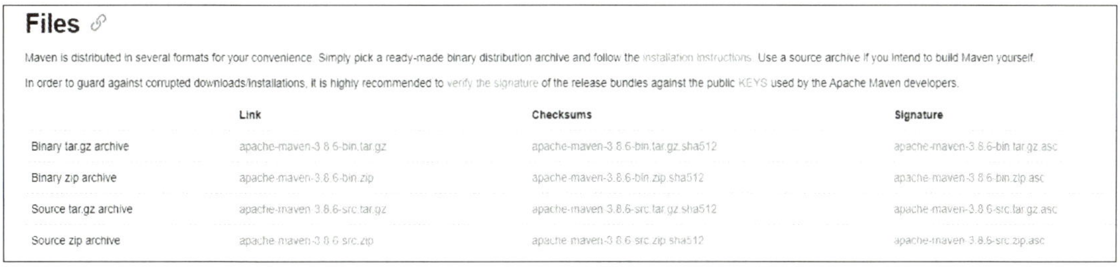

图 1-18 选择版本

将文件解压到存放开发软件的目录，如图 1-19 所示。

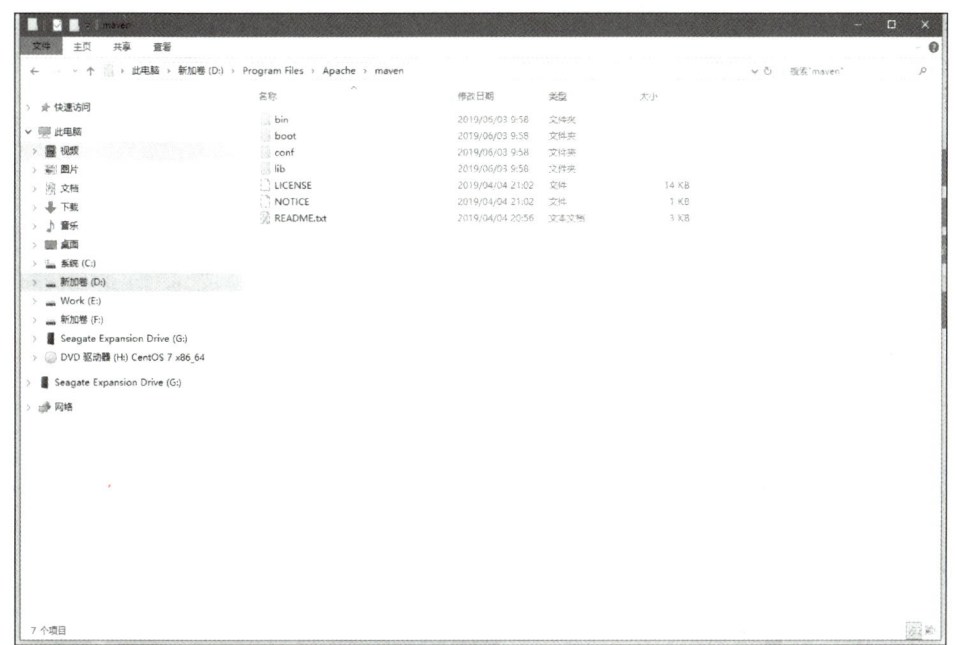

图 1-19 文件解压

新建环境变量 MAVEN_HOME，赋值 D:\Program Files\Apache\maven，如图 1-20 所示。

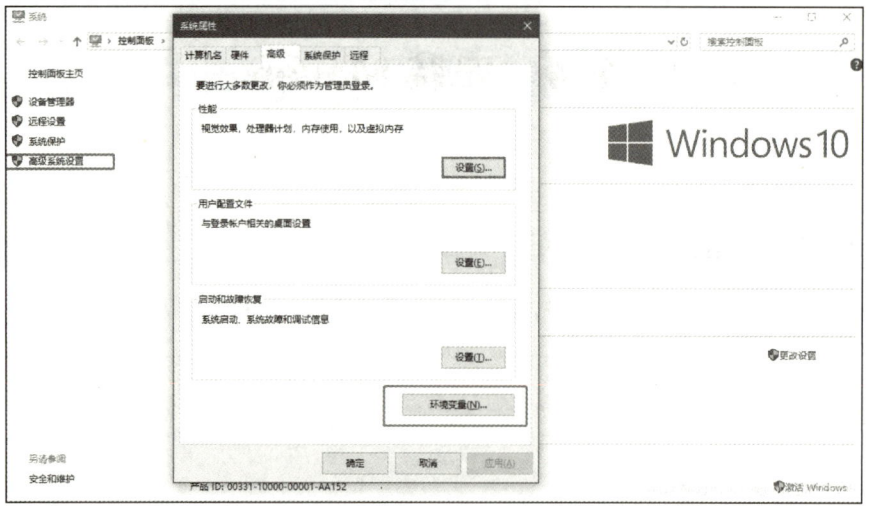

图 1-20　环境变量配置

编辑环境变量 Path，追加%MAVEN_HOME% \ bin，如图 1-21 所示。

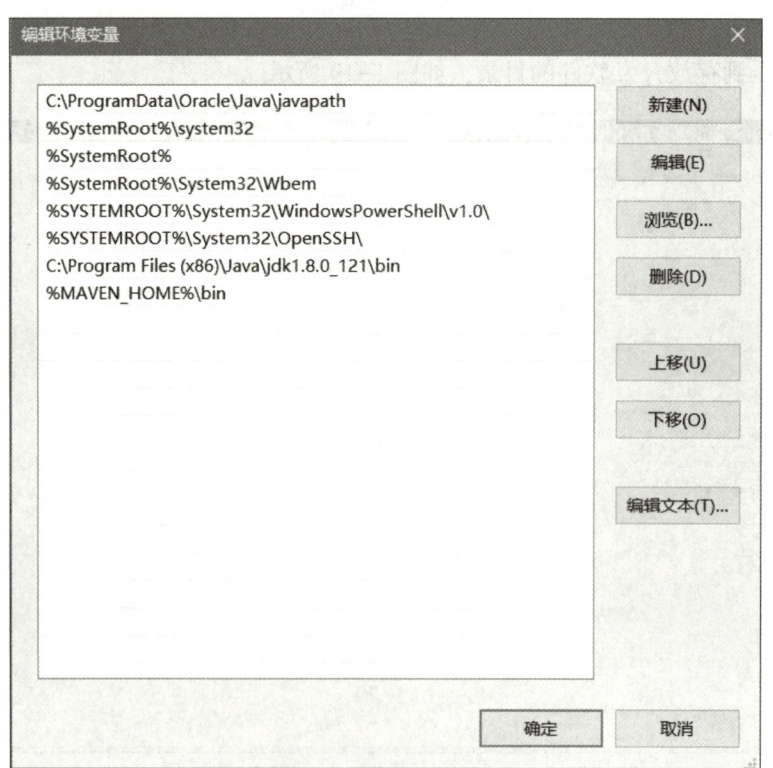

图 1-21　环境变量编辑

至此，Maven 已经完成了安装，我们可以通过 DOS 命令检查一下是否安装成功。测试如图 1-22 所示。

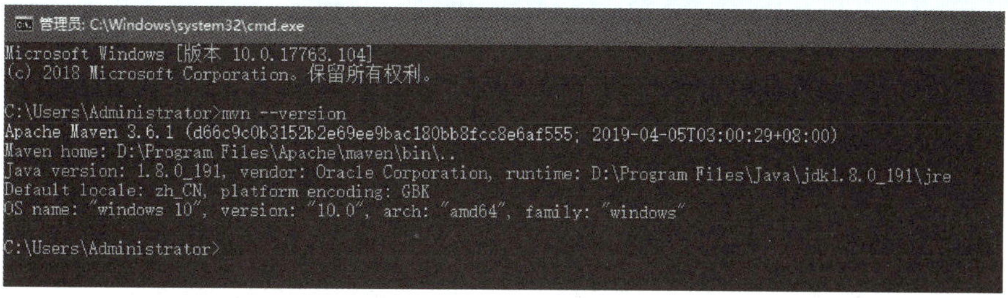

第 1 章　环境介绍

图 1-22　测试

1.4.2　配置 Maven 本地仓库

在 D：\ Program Files \ Apache \（自己选择）目录下新建 maven-repository 文件夹，该目录用作 Maven 的本地库。

打开 D：\ Program Files \ Apache \ maven \ conf \ settings.xml（自己解压 maven 程序的位置下）文件，查找下面这行代码：

```
/path/to/local/repo
```

localRepository 节点默认是被注释掉的，需要把它移到注释之外，然后将 localRepository 节点的值改为我们在前面创建的目录 D：\ ProgramFiles \ Apache \ maven-repository。

localRepository 节点用于配置本地仓库，本地仓库其实起到了一个缓存的作用，它的默认地址是 C：\ Users \ 用户名 .m2。

当我们从 Maven 中获取 jar 包的时候，Maven 首先会在本地仓库中查找，如果本地仓库有则返回，如果没有则从远程仓库中获取包，并在本地仓库中保存。

1.4.3　IDEA 中配置 Maven

首先打开 IDEA 软件的主页，点击 File（文件），选择 Close Project（关闭项目），如图 1-23 所示。

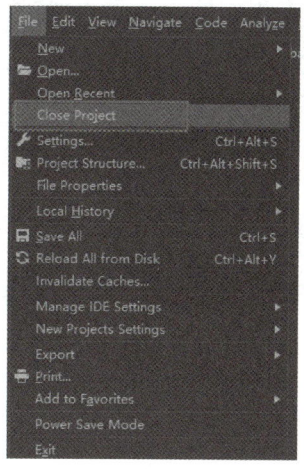

图 1-23　File 菜单

进入到全局配置页，如图 1-24 所示。

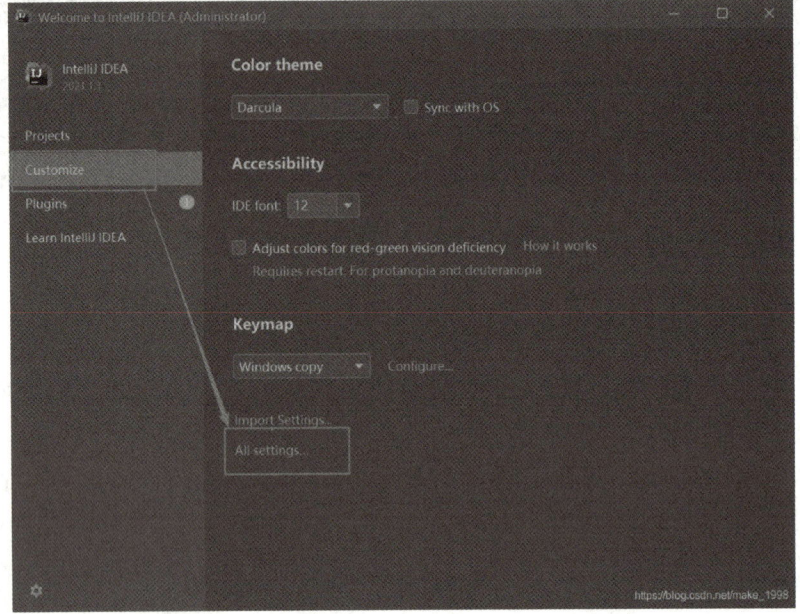

图 1-24　全局配置页

设置 Maven 仓库地址，设置完成后记得点击 OK，如图 1-25 所示。

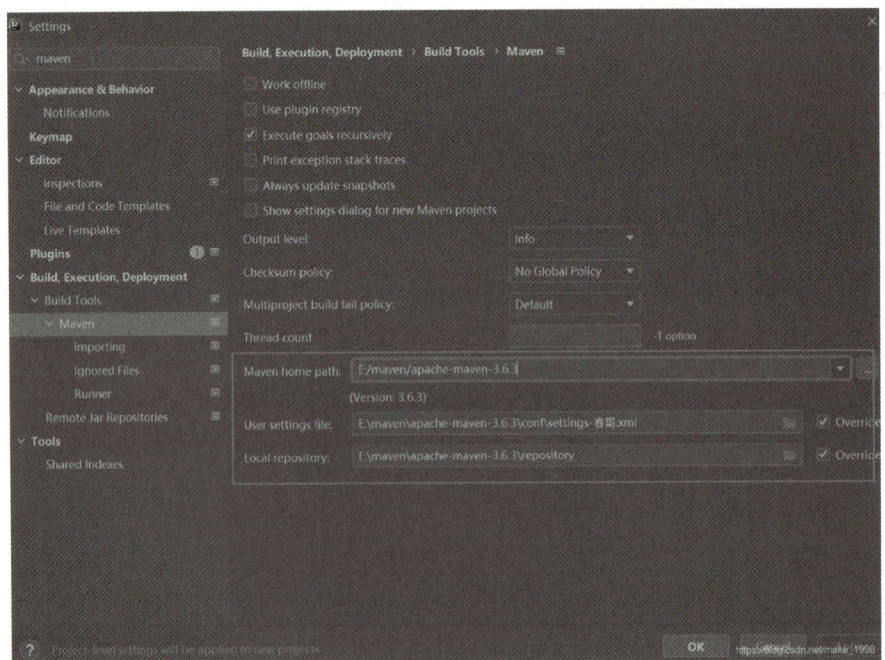

图 1-25　设置 Maven 仓库地址

本章小结

本章主要讲解了本书学习所需要使用到的 JDK、Maven、编码工具 IDEA 的安装步骤以及环境变量的配置,通过本章的学习可以熟练掌握安装与配置开发环境的方法。

课后习题

1. 请简述 Maven 安装步骤。
2. 请简述如何配置 Maven 本地仓库。

第 2 章 初识 MyBatis

[本章知识点]
　　MyBatis 的基本原理及概念
　　MyBtais 入门

[思政目标]
　　使学生树立正确的技能观，努力提高自己的职业技能，为社会和人民造福，绝不利用自己的技能去做违法犯罪之事。
　　培养学生的软件工匠精神，提高学生的综合职业素养，使学生树立社会主义职业精神。

2.1 简　介

　　MyBatis 本是 Apache 的一个开源项目(iBatis)，2010 年这个项目由 Apache Software Foundation 迁移到了 Google Code，并且改名为"MyBatis"，2013 年 11 月迁移到 Github。

　　iBatis 一词来源于"internet"和"abatis"的组合，是一个基于 Java 的持久层框架。iBatis 提供的持久层框架包括 SQL Maps 和 Data Access Objects(DAOs)。

　　MyBatis 是一款优秀的持久层框架，它支持自定义 SQL、存储过程以及高级映射。MyBatis 免除了几乎所有的 JDBC 代码以及设置参数和获取结果集的工作。MyBatis 可以通过简单的 XML 或注解来配置和映射原始类型、接口和 Java POJO(Plain Old Java Objects，普通老式 Java 对象)为数据库中的记录。

2.2 特　点

（1）简单易学：本身就很小且简单，没有任何第三方依赖，最简单的话只要安装两个 jar 文件并配置几个 SQL 映射文件。易于学习，易于使用，通过文档和源代码，就可以较好掌握它的设计思路和实现。

（2）灵活：不会对应用程序或者数据库的现有设计强加任何影响。SQL 写在 XML 里，便于统一管理和优化。通过 SQL 语句可以满足操作数据库的所有需求。

（3）解除了 SQL 与程序代码的耦合：通过提供 DAO 层，将业务逻辑和数据访问逻辑分离，使系统的设计更清晰，更易维护，更易进行单元测试。SQL 和代码的分离，提高了可维护性。

（4）提供映射标签，支持对象与数据库的 orm 字段关系映射。

（5）提供对象关系映射标签，支持对象关系组建维护。

（6）提供 XML 标签，支持编写动态 SQL。

2.3　MyBatis 的工作原理

为了使读者能够更加清晰地理解 MyBatis 程序，在正式讲解 MyBatis 入门案例之前，先来了解一下 MyBatis 程序的工作原理，如图 2-1 所示。

从图 2-1 可以看出，MyBatis 框架在操作数据库时，大体经过了 8 个步骤。下面就对图中的每一步流程进行详细讲解。

①读取 MyBatis 配置文件 mybatis-config.xml。mybatis-config.xml 作为 MyBatis 的全局配置文件，配置了 MyBatis 的运行环境等信息，其中主要内容是获取数据库连接。

②加载映射文件 Mapper.xml。Mapper.xml 文件即 SQL 映射文件，该文件中配置了操作数据库的 SQL 语句，需要在 mybatis-config.xml 中加载才能执行。mybatis-config.xml 可以加载多个配置文件，每个配置文件对应数据库中的一张表。

③构建会话工厂。通过 MyBatis 的环境等配置信息构建会话工厂 SqlSessionFactory。

④创建 SqlSession 对象。由会话工厂创建 SqlSession 对象，该对象中包含了执行 SQL 的所有方法。

⑤MyBatis 底层定义了一个 Executor 接口来操作数据库，它会根据 SqlSession 传递的参数动态地生成需要执行的 SQL 语句，同时负责查询缓存的维护。

⑥在 Executor 接口的执行方法中，包含一个 MappedStatement 类型的参数，该参数是对映射信息的封装，用于存储要映射的 SQL 语句的 id、参数等。Mapper.xml 文件中一个 SQL 对应

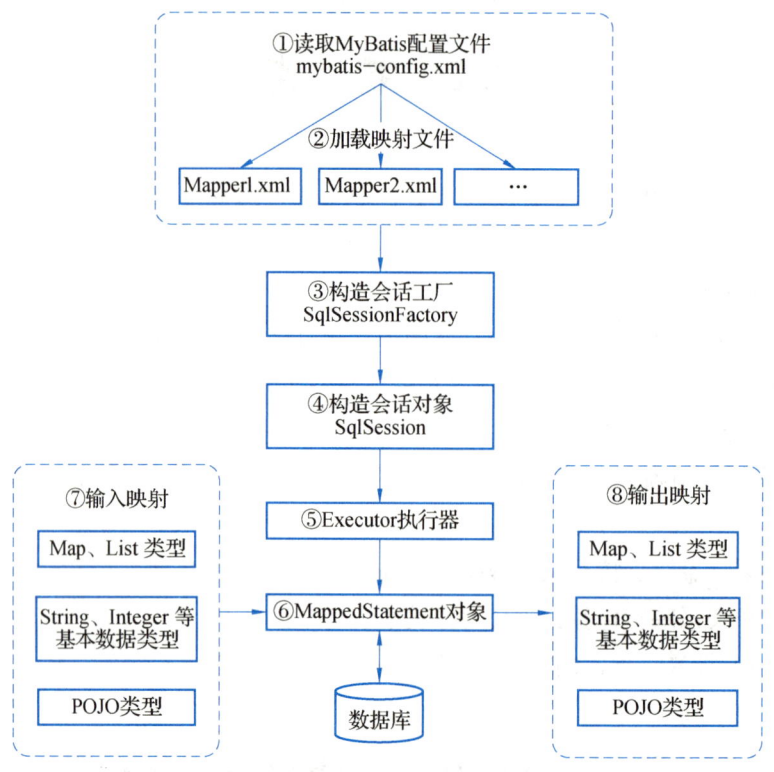

图 2-1　MyBatis 程序的工作原理

一个 MappedStatement 对象，SQL 的 id 即是 MappedStatement 的 id。

⑦输入参数映射。在执行方法时，MappedStatement 对象会对用户执行 SQL 语句的输入参数进行定义（可以定义为 Map 类型、List 类型、基本类型、POJO 类型），Executor 执行器会通过 MappedStatement 对象在执行 SQL 前，将输入的 Java 对象映射到 SQL 语句中。这里对输入参数的映射过程就类似于 JDBC 编程中对 preparedStatement 对象设置参数的过程。

⑧输出结果映射。在数据库中执行完 SQL 语句后，MappedStatement 对象会对 SQL 执行输出的结果进行定义（可以定义为 Map 类型、List 类型、基本类型、POJO 类型），Executor 执行器会通过 MappedStatement 对象在执行 SQL 语句后，将输出结果映射至 Java 对象中。这种将输出结果映射到 Java 对象的过程就类似于 JDBC 编程中对结果的解析处理过程。

通过上面对 MyBatis 框架执行流程的讲解，相信读者对 MyBatis 框架已经有了一个初步的了解。对于初学者来说，上面所讲解的内容可能不能完全理解，现阶段也不要求读者能完全理解，这里讲解 MyBatis 框架的执行过程是为了方便后面程序的学习。在学习完 MyBatis 框架后，读者自然就会明白上面所讲解的内容了。

第 2 章　初识 MyBatis

2.4　入门案例

（1）在 mybatisdb 数据库创建 book 表，表结构如下所示。

```sql
CREATE TABLE book(
bid INT PRIMARY KEY AUTO_INCREMENT,
bname VARCHAR(15) NOT NULL,
btype INT,
price DOUBLE DEFAULT 0.0,
bdesc VARCHAR(100)
);
```

（2）在 IDEA 中创建 Maven 项目，如图 2-2 所示。

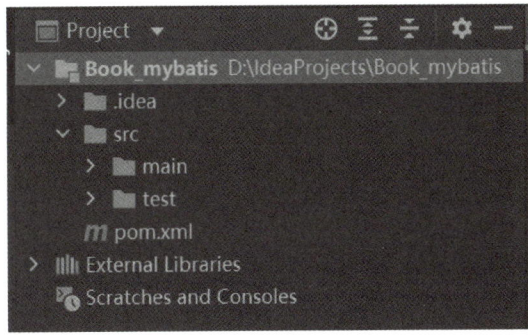

图 2-2　创建 Maven 项目

（3）在 pom 文件中设置依赖 MyBatis 和 MySQL 的 jar 包。

```xml
<?xml version="1.0" encoding="UTF-8"?>
<project xmlns="http://maven.apache.org/POM/4.0.0"
xmlns:xsi="http://www.w3.org/2001/XMLSchema-instance"
xsi:schemaLocation="http://maven.apache.org/POM/4.0.0 http://maven.apache.org/xsd/maven-4.0.0.xsd">
<modelVersion>4.0.0</modelVersion>
<groupId>org.example</groupId>
<artifactId>Book_mybatis</artifactId>
<version>1.0-SNAPSHOT</version>
<properties>
<maven.compiler.source>8</maven.compiler.source>
<maven.compiler.target>8</maven.compiler.target>
</properties>
<dependencies>
<dependency>
<groupId>mysql</groupId>
<artifactId>mysql-connector-java</artifactId>
```

```xml
<version>5.0.8</version>
</dependency>
<!-- https://mvnrepository.com/artifact/log4j/log4j -->
<dependency>
<groupId>log4j</groupId>
<artifactId>log4j</artifactId>
<version>1.2.16</version>
</dependency>
<dependency>
<groupId>org.mybatis</groupId>
<artifactId>mybatis</artifactId>
<version>3.1.0</version>
</dependency>
<dependency>
<groupId>junit</groupId>
<artifactId>junit</artifactId>
<version>4.12</version>
</dependency>
</dependencies>
</project>
```

（4）在 src 的 com.test.bean 包下创建实体类 Book，并且为其设置 getter、setter 以及有参与无参构造方法。

```java
package com.test.bean;
public class Book {
    private Integer bid;
    private String bname;
    private Integer btype;
    private Double price;
    private String bdesc;
    public Book() {
    }
    public Book(Integer bid, String bname, Integer btype, Double price, String bdesc) {
        this.bid = bid;
        this.bname = bname;
        this.btype = btype;
        this.price = price;
        this.bdesc = bdesc;
    }
    public void setBid(Integer bid) {
        this.bid = bid;
    }
    public void setBname(String bname) {
        this.bname = bname;
    }
```

```java
public void setBtype(Integer btype){
this.btype = btype;
}
public void setPrice(Double price){
this.price = price;
}
public void setBdesc(String bdesc){
this.bdesc = bdesc;
}
public Integer getBid(){
return bid;
}
public String getBname(){
return bname;
}
public Integer getBtype(){
return btype;
}
public Double getPrice(){
return price;
}
public String getBdesc(){
return bdesc;
}
}
```

（5）在 resources 下创建 MyBatis 的配置文件 Mybatis-config.xml，该配置文件主要有如下两个重要配置：

① 数据源配置。

② 接口映射文件配置。

```xml
<!DOCTYPE configuration PUBLIC
"-//mybatis.org//DTD Config 3.0//EN"
         "http://mybatis.org/dtd/mybatis-3-config.dtd">
<configuration>
<environments default="mysql">
<environment id="mysql">
<transactionManager type="JDBC"></transactionManager>
<dataSource type="POOLED">
<property name="driver" value="com.mysql.jdbc.Driver"/>
<property name="url" value="jdbc:mysql://localhost:3306/mybatisdb"/>
<property name="username" value="root"/>
<property name="password" value="root"/>
</dataSource>
</environment>
</environments>
</configuration>
```

(6) 在 src 的 com.test.factory 包下创建 MyBatis 的工厂类，主要作用如下：
① 读取 MyBatis 的配置文件，然后构建出 SqlSessionFactory。
② 从 SqlSessionFactory 中打开一个 SqlSession，用于获得数据访问层接口的代理对象。

```java
package com.test.factory;
import org.apache.ibatis.io.Resources;
import org.apache.ibatis.session.SqlSession;
import org.apache.ibatis.session.SqlSessionFactory;
import org.apache.ibatis.session.SqlSessionFactoryBuilder;
import java.io.Reader;
public class MybatisFactory{
    private static SqlSessionFactory ssf;
    private static SqlSession session;
    static{
        try{
            Reader reader = Resources.getResourceAsReader("Mybatis-config.xml");
            ssf = new SqlSessionFactoryBuilder().build(reader);
        }catch(Exception e){
            e.printStackTrace();
        }
    }
    public static SqlSession openSession(){
        session = ssf.openSession();
        return session;
    }
}
```

(7) 在 com.test.dao 包创建数据访问接口 IBookDao。

```java
package com.test.dao;
import com.test.bean.Book;
import java.util.List;
import java.util.Map;
public interface IBookDao{
    public int insertBook(Book book) throws Exception;
    public int deleteBookByBid(Integer bid) throws Exception;
    public int updateBookByBid(Book book) throws Exception;
    public Book queryBookByBid(Integer bid) throws Exception;
    public List<Book> queryAllBook() throws Exception;
    public Integer queryBookCount() throws Exception;
}
```

(8) 在 resources 下为数据访问接口创建 SQL 映射文件（IBookDao-mapper.xml）。

```xml
<!DOCTYPE mapper PUBLIC
    "-//mybatis.org//DTD Mapper 3.0//EN"
        "http://mybatis.org/dtd/mybatis-3-mapper.dtd">
<mapper namespace="com.test.dao.IBookDao">
<insert id="insertBook" parameterType="com.test.bean.Book">
```

```xml
            insert into book(bname,btype,price,bdesc)
            value (#{bname},#{btype},#{price},#{bdesc});
</insert>
<delete id="deleteBookByBid" parameterType="int">
            delete from book where bid=#{bid}
</delete>
<update id="updateBookByBid" parameterType="com.test.bean.Book">
            update book set
                bname=#{bname},btype=#{btype},
                price=#{price},bdesc=#{bdesc}
where
bid=#{bid}
</update>
<select id="queryBookByBid" parameterType="int" resultType="com.test.bean.Book">
            select * from book where bid=#{bid}
</select>
<select id="queryAllBook"    resultType="com.test.bean.Book">
            select bid,bname,btype,price,bdesc from book
</select>
<select id="queryBookCount"    resultType="int">
            select count(*) from book
</select>
</mapper>
```

(9) 将接口映射文件配置到 MyBatis 的配置文件中，位置如图 2-3 所示。

```xml
<mappers>
<mapper resource="IBookDao-mapper.xml"></mapper>
</mappers>
```

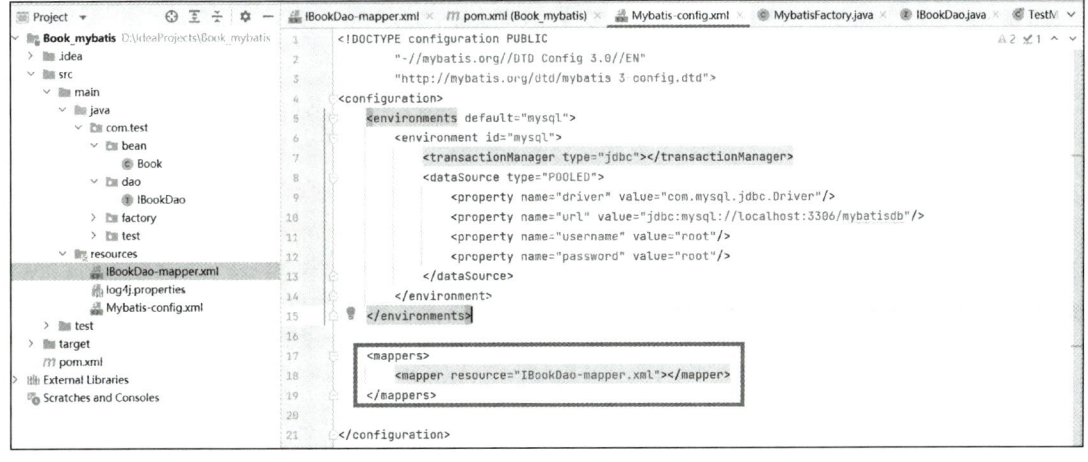

图 2-3　MyBatis 的配置文件

(10)编写测试类对前面所做的数据访问层接口进行测试。

①测试添加。

```java
package com.test.test;
import com.test.bean.Book;
import com.test.dao.IBookDao;
import com.test.factory.MyBatisFactory;
import org.apache.ibatis.session.SqlSession;
import org.junit.After;
import org.junit.Before;
import org.junit.Test;
import java.util.ArrayList;
import java.util.HashMap;
import java.util.List;
import java.util.Map;
public class TestMain{
    IBookDao dao=null;
    SqlSession session=null;
    @Before
    public void before(){
        session= MyBatisFactory.openSession();
        dao=session.getMapper(IBookDao.class);
    }
    @After
    public void after(){
        session.commit();
        session.close();
    }
    @Test
    public void testInsert() throws Exception{
        Book book=new Book(null,"十万个为什么",1,59.9,"没有啥简介");
        int i=dao.insertBook(book);
        System.out.println(i);
    }
}
```

②测试修改。

```java
@Test
public void testUpdate() throws Exception{
    Book book=new Book(2,"十万个为什么",2,60.0,"没有啥简介");
    int i=dao.updateBookByBid(book);
    System.out.println(i);
}
```

③测试删除。

```java
@Test
public void testDelete() throws Exception{
```

```
int i=dao.deleteBookByBid(3);
System.out.println(i);
}
```

④测试根据 id 查询。

```
@Test
public void testQueryById() throws Exception {
Book book=dao.queryBookByBid(2);
System.out.println(book);
}
```

⑤测试查询所有。

```
@Test
public void testQueryAll() throws Exception {
List<Book>list=dao.queryAllBook();
list.forEach(book->System.out.println(book));
}
```

⑥测试查询总行数。

```
@Test
public void testQueryBookCount() throws Exception {
int count=dao.queryBookCount();
System.out.println(count);
}
```

本章小结

本章首先对 MyBatis 框架的概念、特点进行了讲解，然后对 MyBatis 框架的工作原理进行了流程分析，最后通过一个简单的增删改查案例来演示 MyBatis 框架的基本使用。通过本章的学习，读者可以了解 MyBatis 的概念和作用，熟悉 MyBatis 的工作原理，并能够使用 MyBatis 框架完成基本的数据库操作。

课后习题

1. 请简述 MyBatis 框架与 Hibernate 框架的区别。
2. 请简述 MyBatis 的工作执行流程。

第 3 章
MyBatis 的核心配置

  学习目标

[本章知识点]

配置文件的主要元素介绍：

1. <environments>元素

2. <mappers>元素

核心类介绍：

1. SqlSessionFactory

2. SqlSession

映射文件的主要标签介绍：

1. <select>标签

2. <insert>标签

3. <update>标签

4. <delete>标签

5. <sql>标签

6. <resultMap>标签

[思政目标]

让学生了解 SSM 框架开发规范的重要性，培养学生的职业素质和道德规范。

通过上一章的学习，读者对 MyBatis 框架的使用已经有了一个初步的了解，但是要想熟练地使用 MyBatis 框架进行实际开发，只会简单的配置是不行的，还需要对框架中的核心对象以及映射文件和配置文件有更加深入的了解。接下来，本章将对这些内容进行详细的讲解。

3.1　配置文件的主要元素介绍

```
<!DOCTYPE configuration PUBLIC
"-//mybatis.org//DTD Config 3.0//EN"
    "http://mybatis.org/dtd/mybatis-3-config.dtd">
<configuration>
<environments default="mysql">
<environment id="mysql">
<transactionManager type="JDBC"></transactionManager>
<dataSource type="POOLED">
<property name="driver" value="com.mysql.jdbc.Driver"/>
<property name="url" value="jdbc:mysql://localhost:3306/mybatisdb"/>
<property name="username" value="root"/>
<property name="password" value="root"/>
</dataSource>
</environment>
</environments>
<mappers>
<mapper resource="IBookDao-mapper.xml"></mapper>
</mappers>
</configuration>
```

在 MyBatis 配置文件中有如下两个主要的元素。

（1）<environments>元素。

该元素用于配置数据源环境。MyBatis 可以配置多种环境，属性 default 指定使用某种环境。

<environment>配置一个具体的环境信息，必须有 id 属性和<transactionManager>元素。id 代表当前环境的唯一标识；<transactionManager>是事务管理器，它的 type 属性指定事务管理器的类型，取值如下：

JDBC（JdbcTransactionFactory）

MANAGED（ManagedTransactionFactory）

<dataSource>配置数据源，它的 type 属性指定数据源类型，取值如下：

UNPOOLED（UnpooledDataSourceFactory）

POOLED（PooledDataSourceFactory）

JNDI（JndiDataSourceFactory）

（2）<mappers>元素。

<mappers>元素下有许多<mapper>元素，每一个<mapper>元素中配置的都是一个独立的

映射配置文件的路径,配置方式有以下几种。

第一种:使用相对路径进行配置。示例代码如下。

```xml
<mappers>
<mapper resource="org/mybatis/mappers/UserMapper.xml"/>
<mapper resource="org/mybatis/mappers/ProductMapper.xml"/>
<mapper resource="org/mybatis/mappers/ManagerMapper.xml"/>
</mappers>
```

第二种:使用绝对路径进行配置。示例代码如下。

```xml
<mappers>
<mapper url="file:///var/mappers/UserMapper.xml"/>
<mapper url="file:///var/mappers/ProductMapper.xml"/>
<mapper url="file:///var/mappers/ManagerMapper.xml"/>
</mappers>
```

第三种:使用接口信息进行配置。示例代码如下。

```xml
<mappers>
<mapper class="org.mybatis.mappers.UserMapper"/>
<mapper class="org.mybatis.mappers.ProductMapper"/>
<mapper class="org.mybatis.mappers.ManagerMapper"/>
</mappers>
```

第四种:使用接口所在包进行配置。示例代码如下。

```xml
<mappers>
<package name="org.mybatis.mappers"/>
</mappers>
```

只有配置了 mappers 信息后,MyBatis 才知道去哪里加载 Mapper 映射文件。在日常开发中,可以根据项目中 Mapper 的配置偏好,选择适合配置文件的配置方式。

3.2 核心类介绍

```java
package com.test.factory;
import org.apache.ibatis.io.Resources;
import org.apache.ibatis.session.SqlSession;
import org.apache.ibatis.session.SqlSessionFactory;
import org.apache.ibatis.session.SqlSessionFactoryBuilder;
import java.io.Reader;
public class MyBatisFactory{
    private static SqlSessionFactory ssf;
    private static SqlSession session;
```

```
static {
try {
Reader reader = Resources.getResourceAsReader("MyBatis-config.xml");
ssf = new SqlSessionFactoryBuilder().build(reader);
    } catch (Exception e) {
e.printStackTrace();
        }
    }
public static SqlSessionopenSession() {
session = ssf.openSession();
return    session;
    }
}
```

3.2.1　SqlSessionFactory

SqlSessionFactory 是通过 SqlSessionFactoryBuilder 的 build 方法创建的，build 方法创建的是 SqlSessionFactory 的实现类 DefaultSqlSessionFactory，这个实现类主要用的设计模式是建造者（build）模式。SqlSessionFactory 要创建出 DefaultSqlSession，这个是 SqlSession 的实现类。

3.2.2　SqlSession

SqlSession 与 SqlSessionFactory 是同一个接口，由 SqlSessionFactory.openSession() 创建的 SqlSession 相当于 DefaultSqlSession 这个实现类。DefaultSqlSession 这个实现类通过调用 Executor 执行器可以进行增删查改等操作。

平时 SqlSession.select() 等操作其实是在操作 DefaultSqlSession.select() 的方法，而 DefaultSqlSession 的方法其实也不是直接对数据库进行操作的，而是通过调用 Executor 执行器来操作，而 Executor 执行器其实也是一个接口。在解析配置文件的时候已经解析出来 SQL，那么根据一路传过来的 SQL 以及参数等信息，Executor 再调度 StatementHandler 等对象对数据库进行增删查改操作。

3.3 映射文件的主要标签介绍

3.3.1 <select>标签

```
<select id="queryBookByBid" parameterType="int" resultType="com.test.bean.Book">
    select * from book where bid=#{bid}
</select>
<select id="queryAllBook"  resultType="com.test.bean.Book">
    select <include refid="columns_sql"/> from book
</select>
<select id="queryBookCount"  resultType="int">
    select count(*) from book
</select>
```

<select>是用于编写查询语句用的标签，下面为<select>标签常用属性介绍。

id：表示当前<select>标签的唯一标识，指向接口中的方法名。

parameterType：指定查询限制条件的输入类型，一般使用#{}实现的是向 prepareStatement 中的预处理语句中设置参数值。

resultType：指定查询返回结果的输出类型，如果返回的结果是一个实体类，必须要求实体类的属性和表的字段名称相同。

resultMap：也是一个输出类型，配合<resultMap>标签使用。

flushCache：设置查询的时候是否清空缓存，默认为 false。

useCache：将查询结果放入缓存中，默认为 true。

timeout：设置查询返回结果的最大响应时间。

fetchSize：每次批量返回的结果行数，默认不设置。

注意：<select>标签必须设置返回结果类型。

3.2.2 <insert>标签

```
<insert id="insertBook"
parameterType="com.test.bean.Book">
    insert into book(bname,btype,price,bdesc)
value (#{bname},#{btype},#{price},#{bdesc});
</insert>
```

用于编写插入语句用的标签，需要注意的是，<insert>标签不允许设置返回类型，默认返回受影响行数。

3.2.3 \<update>标签

```
<update id="updateBookByBid" parameterType="com.test.bean.Book">
    update book set
bname=#{bname},btype=#{btype},
price=#{price},bdesc=#{bdesc}
where bid=#{bid}
</update>
```

用于编写修改语句用的标签,需要注意的是,update 标签不允许设置返回类型,默认返回受影响行数。

3.2.4 \<delete>标签

```
<delete id="deleteBookByBid" parameterType="int">
    delete from book
where bid=#{bid}
</delete>
```

用于编写删除语句用的标签,需要注意的是,delete 标签不允许设置返回类型,默认返回受影响行数。

3.2.5 \<sql>标签

```
<sql id="columns_sql">
bname,btype,price,bdesc
</sql>
<select id="queryAllBook"
resultType="com.test.bean.Book">
    select <include refid="columns_sql"/> from book
</select>
```

可以重用的 SQL 语句,可以通过 include 标签被其他语句引用。

3.2.6 \<resultMap>标签

```
<resultMap id="rm_book" type="com.test.bean.Book">
<result property="bid" column="bid"></result>
<result property="bname" column="bname"></result>
```

```
<result property="btype" column="btype"></result>
<result property="price" column="price"></result>
<result property="bdesc" column="bdesc"></result>
</resultMap>
```

<resultMap>用于解决实体类中属性和表字段名不相同的问题。
id：表示当前<resultMap>标签的唯一标识。
type：表示当前<resultMap>标签配置的是哪一个实体类映射关系。
<result>用于定义表字段和实体类属性的对应关系。
property：记录实体类的属性。
column：记录表的字段名称。

3.2.7 <mapper>标签

```
<mapper namespace="com.test.dao.IBookDao">
</mapper>
```

<mapper>指每个映射文件的根标签。要重点关注<mapper>标签中的 namespace 属性。namespace 指向的是 dao 的接口，表示为哪一个接口创建 SQL 映射。

本章小结

本章主要对 MyBatis 中的核心配置文件和映射文件常用标签进行了讲解，对两个核心类 SqlSessionFactory 与 SqlSession 进行了讲解。首先讲解了配置文件中的常用标签及其使用，然后介绍 MyBatis 中的两个重要核心对象 SqlSessionFactory 和 SqlSession，最后对映射文件中的几个主要元素进行了详细讲解。通过本章的学习，读者将能够了解 MyBatis 中两个核心对象的作用，熟悉配置文件中常用标签的使用，并掌握映射文件中常用元素的使用。

课后习题

1. 请简述 MyBatis 核心对象 SqlSessionFactory 的获取方式。
2. 请简述 MyBatis 映射文件中的主要元素及其作用。

第4章 动态 SQL

学习目标

[本章知识点]

动态 SQL 中的标签介绍：
<if>元素
<choose>、<when>、<otherwise>元素
<where>、<trim>元素
<foreach>元素

[思政目标]

培养学生树立正确的技能观，努力提高自己的技能，为社会和人民造福。
培养学生团结协作、合作共赢的意识。

开发人员在使用 JDBC 或其他类似的框架进行数据库开发时，通常都要根据需求去手动拼装 SQL，这是一个非常麻烦且痛苦的工作，而 MyBatis 提供的对 SQL 语句动态组装的功能，恰能很好地解决这一问题。在本章中，我们将对 MyBatis 框架的动态 SQL 进行详细讲解。

4.1　动态 SQL 介绍

MyBatis 的强大特性之一便是它的动态 SQL。如果前期有使用 JDBC 或其他类似框架的经验，读者就能体会到根据不同条件拼接 SQL 语句的困难和麻烦：拼接的时候要确保不能忘了必要的空格，还要注意省掉列名列表最后的逗号。利用动态 SQL 这一特性可以彻底摆脱这种困难和麻烦。

4.2　动态 SQL 标签介绍

动态 SQL 标签	说明
<if>	判断语句，用于单条件分支判断
<choose>、(<when>、<otherwise>)	相当于 Java 中的 switch…case…default 语句，用于多条件分支判断
<where>、<trim>	辅助元素，用于处理一些 SQL 拼装，以及特殊字符问题
<foreach>	循环语句，常用于 in 语句等列举条件中

4.3　动态 SQL 标签案例

4.3.1　<where>、<if>应用

根据书籍名称以及书籍类型查询书籍。

```
/**
* 根据书籍名称以及书籍类型查询书籍
* @param name
* @param type
* @return
*/
public List<Book>queryBookByNameAndType(@Param("name") String name, @Param("type") Integer type);
```

注意，当接口方法需要传入多个参数(大于 1 个参数)到 SQL 时，有如下几种方式：
①以 bean 的形式传入，通过成员属性取值，参考修改方法以及添加方法。
②以 map 的形式传入，通过 key 取值。
③以注解@Param("")传入，通过注解括号中指定的参数名取值。

```xml
<select id="queryBookByNameAndType"
    resultType="com.test.bean.Book"
    parameterType="map">
        select * from book
<where>
<if test="name!=null and name!=''">
bname=#{name}
</if>
<if test="type!=null and type!=0">
        and btype=#{type}
</if>
</where>

</select>
```

从上方代码可以看出，在编写 queryBookByNameAndType 方法的映射 SQL 时，使用到了 <where>以及<if>标签。

使用<where>标签，当 where 标签中有内容时，会在 SQL 语句后自动添加 where 关键字，反之则不会添加 where 关键字。

而使用<if>标签时，相当于有一个判断效果，test 属性为判断条件，test 中间可以判断传入的参数，从而动态生成 SQL 语句。

4.3.2 \<choose>、\<when>、\<otherwise>应用

根据书籍 id 修改书籍类型。

```java
/**
* 根据书籍 id 修改书籍类型
* @param id
* @param type
* @return
*/
public int updateBookType(@Param("id") Integer id,
                         @Param("type") Integer type);
```

```xml
<update id="updateBookType" parameterType="map">
    update book set
<choose>
<when test="type==1">
btype=1
</when>
<when test="type==2">
btype=2
</when>
<when test="type==3">
```

— 33 —

```
            btype = 3
        </when>
        <otherwise>
            btype = 0
        </otherwise>
    </choose>
        where bid = #{id}
</update>
```

如上代码在完成 updateBookType 方法的 SQL 映射时，使用到<choose>、<when>、<otherwise>三个动态 SQL 标签，这三个标签类似于 Java 语法中的 switch 语法。

<choose>会从子标签<when>以及<otherwise>中选取一个成立的拼接出完整的 SQL 语句。

<when>用于判断 test 条件是否成立，成立则选取该内容进行 SQL 拼接。

<otherwise>作用为当 when 都未成立时，使用该默认内容进行 SQL 拼接。

4.3.3 \<trim>应用

根据条件查询书籍信息。

```
/**
 * 根据条件查询书籍信息
 * @param map {bid,bname,btype,price,bdesc}
 * @return
 */
public List<Book>selectBookByParam(Map map);

<select id="selectBookByParam" parameterType="map"
resultType="com.test.bean.Book">
    select * from book
<trim prefix="where" prefixOverrides="and | or ">
<if test="bid != null">
            and bid=#{bid}
</if>
<if test="bname != null">
            and bname=#{bname}
</if>
<if test="btype != null">
            and  btype=#{btype}
</if>
<if test="price != null">
            and  price=#{price}
</if>
<if test="bdesc != null">
            and  bdesc like '%${bdesc}%'
</if>
```

```
</trim>
</select>
```

prefixOverrides 属性会忽略通过管道分隔的文本序列(注意此例中的空格也是必要的)。它带来的结果就是所有在 prefixOverrides 属性中指定的内容将被移除，并且插入 prefix 属性中指定的内容。

注意 $与#传值区别：

①#{}是预编译处理，${}是字符串替换(当作占位符来用)。

②MyBatis 在处理#{}时，会将 SQL 中的#{}替换为？号，调用 PreparedStatement 的 set 方法来赋值。

③MyBatis 在处理 ${}时，就是把 ${}替换成变量的值。

④使用#{}可以有效地防止 SQL 注入，提高系统安全性。

⑤在某些特殊场合下只能用 ${}，不能用#{}。例如：在使用排序时(ORDER BY ${id})，如果使用#{id}，则会被解析成 ORDER BY "id"，这显然是一种错误的写法。

4.3.4 \<foreach>应用

根据 id 批量查询。

```
/**
* 根据 id 批量查询
* select * from book where id in (1,2,3,4);
* @ param ids
* @ return
*/
public List<Book>selectBookByIds(@ Param("ids") List ids);

<select id = "selectBookByIds" parameterType = "map"
                resultType = "com.test.bean.Book">
    select * from book where bid in
<foreach collection = "ids"
                item = "id"
                open = "(" close = ")"
                separator = ",">
#{id}
</foreach>
</select>
```

如上代码在完成 selectBookByIds 方法的 SQL 批量查询映射，其中使用到了\<foreach>标签，该标签可以用于遍历传入的参数。

collection：指向需要遍历的集合参数。

item：表示一个临时变量。

open：表示最左边使用什么符号包裹。

close：表示最右边使用什么符号包裹。

separator：表示每个遍历出来的变量使用什么符号分隔。

本章小结

本章首先对 MyBatis 框架的动态 SQL 元素作了简要介绍，然后分别对这些主要的动态 SQL 元素进行了详细讲解。通过本章的学习，读者可以了解常用动态 SQL 元素的主要作用，并能够掌握这些元素在实际开发中如何使用。在 MyBatis 框架中，这些动态 SQL 元素的使用十分重要，熟练地使用它们能够极大地提高开发效率。

课后习题

1. 请简述 MyBatis 框架动态 SQL 中的主要元素及其作用。
2. 请简述 MyBatis 框架动态 SQL 中<foreach>元素 collection 属性的注意事项。

第 5 章
MyBatis 的关联关系

[本章知识点]
    关联关系概述
    MyBatis 中的一对一关联关系
    MyBatis 中的一对多关联关系
[思政目标]
    培养学生诚实、守信、坚忍不拔的性格。
    提高学生在沟通表达、自我学习和团队协作方面的能力。

通过前面几章的学习，读者已经熟悉了 MyBatis 的基本知识，并能够使用 MyBatis 以及面向对象的方式进行数据库操作，但这些操作只是针对单表实现的。在实际的开发中，对数据库的操作常常会涉及多张表，这在面向对象中就涉及了对象与对象之间的关联关系。针对多表之间的操作，MyBatis 提供了关联映射，通过关联映射就可以很好地处理对象与对象之间的关联关系。本章将对 MyBatis 的关联关系映射进行详细讲解。

5.1 关联关系概述

在关系型数据库中，多表之间存在着三种关联关系，分别为一对一、一对多和多对多，如图 5-1 所示。

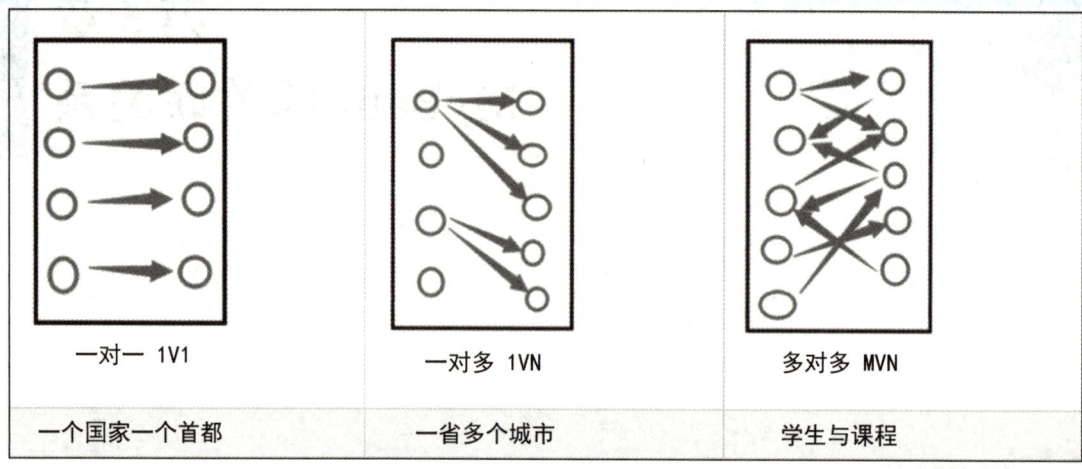

图 5-1 关联关系

这三种关联关系的具体说明如下。
- 一对一：在任意一方引入对方主键作为外键。
- 一对多：在"多"的一方，添加"一"的一方的主键作为外键。
- 多对多：产生中间关系表，引入两张表的主键作为外键，两个主键成为联合主键或使用新的字段作为主键。

通过数据库中的表可以描述数据之间的关系，同样在 Java 中，通过对象也可以进行关系描述，如表 5-1 所示。

表 5-1 数据之间的关系

class A{
　　B b;
}
class B{
　　A a;
}
	class A{
	　　List\<B\> b;
	}
	class B{
	　　A a;
	}
		class A{
		　　List\<B\> b;
		}
		class B{
		　　List\<A\> a;
		}

一对一	一对多	多对多

表 5-1 中的三种关联关系的描述如下。

一对一的关系：在本类中定义对方类型的对象，如 A 类中定义 B 类型的属性 b，B 类中定义 A 类型的属性 a。

一对多的关系：一个 A 类型对应多个 B 类型的情况，需要在 A 类中以集合的方式引入 B 类型的对象，在 B 类中定义 A 类型的属性 a。

多对多的关系：在 A 类中定义 B 类型的集合，在 B 类中定义 A 类型的集合。

以上就是 Java 对象中，三种实体类之间的关联关系。

在 MyBatis 中并没有多对多的实现，这是因为多对多级联可以通过两个一对多级联进行

第 5 章　MyBatis 的关联关系

替换。在接下来的 2 个小节中，将对 MyBatis 中的一对一和多对一关联关系的使用进行详细讲解。

5.2　一对一

在现实生活中，一对一关联关系是十分常见的。例如，登录信息对应一个人的详情信息，同时一个人的详情信息也只会对应一个登录信息。

代码实现如下。

（1）在 mybatisdb 创建登录信息表（tab_login_info）以及用户详情信息表（tab_user_info），如下所示。

```
USE mybatisdb；
CREATE TABLE tab_login_info(
    id INT PRIMARY KEY AUTO_INCREMENT,
    username VARCHAR(11) NOT NULL UNIQUE,
    password VARCHAR(11) NOT NULL,
    uid INT NOT NULL
);
CREATE TABLE tab_user_info(
    uid INT PRIMARY KEY AUTO_INCREMENT,
    name VARCHAR(8) NOT NULL UNIQUE,
    gender INT,
    address VARCHAR(50)
);
```

在表中根据表的结构，输入每个字段的值，如图 5-2 所示。

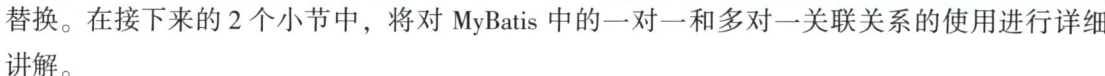

图 5-2　表格输入

（2）创建 Java 工程（CascadeRelationsDemo），并且依赖 MyBatis 以及 MySQL 的 jar 包，如下所示。

```
<?xml version="1.0" encoding="UTF-8"?>
<project xmlns="http://maven.apache.org/POM/4.0.0"
xmlns:xsi="http://www.w3.org/2001/XMLSchema-instance"
xsi:schemaLocation="http://maven.apache.org/POM/4.0.0 http://maven.apache.org/xsd/maven-4.0.0.xsd">
```

— 39 —

```xml
<modelVersion>4.0.0</modelVersion>
<groupId>org.test</groupId>
<artifactId>CascadeRelationsDemo</artifactId>
<version>1.0-SNAPSHOT</version>
<properties>
    <maven.compiler.source>8</maven.compiler.source>
    <maven.compiler.target>8</maven.compiler.target>
</properties>
<dependencies>
    <dependency>
        <groupId>mysql</groupId>
        <artifactId>mysql-connector-java</artifactId>
        <version>5.0.8</version>
    </dependency>
    <!-- https://mvnrepository.com/artifact/log4j/log4j -->
    <dependency>
        <groupId>log4j</groupId>
        <artifactId>log4j</artifactId>
        <version>1.2.16</version>
    </dependency>
    <dependency>
        <groupId>org.mybatis</groupId>
        <artifactId>mybatis</artifactId>
        <version>3.1.0</version>
    </dependency>
    <dependency>
        <groupId>junit</groupId>
        <artifactId>junit</artifactId>
        <version>4.12</version>
    </dependency>
</dependencies>
</project>
```

（3）在 resources 创建 MyBatis 配置文件（mybatis-config.xml），如下所示。

```xml
<!DOCTYPE configuration PUBLIC
"-//mybatis.org//DTD Config 3.0//EN"
        "http://mybatis.org/dtd/mybatis-3-config.dtd">
<configuration>
<environments default="mysql">
<environment id="mysql">
<transactionManager type="jdbc">
</transactionManager>
<dataSource type="POOLED">
<property name="driver" value="com.mysql.jdbc.Driver"/>
<property name="url"
        value="jdbc:mysql://localhost:3306/mybatisdb"/>
```

```
        <property name="username" value="root"/>
        <property name="password" value="root"/>
      </dataSource>
    </environment>
  </environments>
  <mappers>
  </mappers>
</configuration>
```

(4) 创建 MyBatis 的工厂类 MyBatisFactory.java,如下所示。

```
package com.test.factory;
import org.apache.ibatis.io.Resources;
import org.apache.ibatis.session.SqlSession;
import org.apache.ibatis.session.SqlSessionFactory;
import org.apache.ibatis.session.SqlSessionFactoryBuilder;
import java.io.Reader;
public class MyBatisFactory{
    private static SqlSessionFactory ssf;
    private static SqlSession session;
    static{
        try{
            Reader reader = Resources.getResourceAsReader("MyBatis-config.xml");
            ssf = new SqlSessionFactoryBuilder().build(reader);
        }catch(Exception e){
            e.printStackTrace();
        }
    }
    public static SqlSession openSession(){
        session = ssf.openSession();
        return session;
    }
}
```

(5) 创建 LoginInfo 与 UserInfo 的实体类,如下所示。

```
package com.test.bean;
public class UserInfo{
    private Integer uid;
    private String name;
    private Integer gender;
    private String address;
    public Integer getUid(){
        return uid;
    }
    public void setUid(Integer uid){
        this.uid = uid;
    }
    public String getName(){
```

```java
        return name;
    }
    public void setName(String name) {
        this.name = name;
    }
    public Integer getGender() {
        return gender;
    }
    public void setGender(Integer gender) {
        this.gender = gender;
    }
    public String getAddress() {
        return address;
    }
    public void setAddress(String address) {
        this.address = address;
    }
}

package com.test.bean;
public class LoginInfo{
    private Integer id;
    private String username;
    private String password;
    private Integer uid;
    /*
     * 在一的一方定义需要级联查询出来的信息
     * LoginInfo ---- UserInfo
     **/
    private UserInfo userInfo;
    public UserInfo getUserInfo() {
        return userInfo;
    }
    public void setUserInfo(UserInfo userInfo) {
        this.userInfo = userInfo;
    }
    public Integer getId() {
        return id;
    }
    public void setId(Integer id) {
        this.id = id;
    }
    public String getUsername() {
        return username;
    }
    public void setUsername(String username) {
        this.username = username;
```

```
        }
    public String getPassword() {
        return password;
    }
    public void setPassword(String password) {
        this.password = password;
    }
    public Integer getUid() {
        return uid;
    }
    public void setUid(Integer uid) {
        this.uid = uid;
    }
}
```

一个登录信息会包含一个用户信息,所以在登录信息实体类中定义用户信息属性,用于在登录时存储级联查询出来的用户信息。

(6)创建数据访问层 IUserInfoDao 以及其对应的映射文件,如下所示。

```
package com.test.dao;
import com.test.bean.UserInfo;
public interface IUserInfoDao{
    public UserInfo queryUserInfoByUid(Integer uid);
}
```

(7)创建 IUserInfoDao-mapper.xml,如下所示。

```
<!DOCTYPE mapper PUBLIC
"-//mybatis.org//DTD Mapper 3.0//EN"
"http://mybatis.org/dtd/mybatis-3-mapper.dtd">
<mapper namespace="com.test.dao.IUserInfoDao">
    <select id="queryUserInfoByUid" parameterType="int" resultType="com.test.bean.UserInfo">
        select * from tab_user_info where uid=#{uid}
    </select>
</mapper>
```

(8)创建数据访问层 ILoginInfoDao 以及其对应的映射文件 ILoginInfoDaoDao-mapper.xml,如下所示。

```
package com.test.dao;
import com.test.bean.LoginInfo;
import org.apache.ibatis.annotations.Param;
public interface ILoginInfoDao{
    public LoginInfo login(
        @Param("username") String username,
        @Param("password") String password);
}
```

映射文件如下所示。

```xml
<!DOCTYPE mapper PUBLIC
"-//mybatis.org//DTD Mapper 3.0//EN"
        "http://mybatis.org/dtd/mybatis-3-mapper.dtd">
<mapper namespace="com.test.dao.ILoginInfoDao">
<!--
定义 resultMap,手动封装 LoginInfo
    -->
<resultMap id="rm_loginInfo" type="com.test.bean.LoginInfo">
<result property="id" column="id"></result>
<result property="username" column="username"></result>
<result property="password" column="password"></result>
<result property="uid" column="uid"></result>
<association
property="userInfo"
column="uid"
select="com.test.dao.IUserInfoDao.queryUserInfoByUid">
</association>
</resultMap>
<!--
因在查询时需要级联查询出 UserInfo 信息,并且将级联查询出来的信息封装到 LoginInfo 中,直接使用
resultType 无法满足,所以需要使用 resultMap 手动封装
    -->
<select id="login" parameterType="map"
resultMap="rm_loginInfo">
        select * from tab_login_info
where username=#{username} and password=#{password}
</select>
</mapper>
```

在登录时,查询出登录信息后,需要根据登录信息中的 uid 字段再次调用用户信息,根据 uid 查询的方法去查询用户信息,所以在用户信息中定义根据 uid 查询用户信息的方法即可。

因在查询时需要级联查询出 UserInfo 信息,并且将级联查询出来的信息封装到 LoginInfo 中,如图 5-3 所示。但直接使用 resultType 无法满足,所以需要使用 resultMap 手动封装。

定义 resultMap 时需要在 resultMap 中用到 <association> 标签进行级联查询操作。<association>标签中属性功能如下。

property:级联查询出来的内容,指向实体类中一个属性。

column:级联查询时的条件,指向字段名。

select:使用数据访问层中的哪个方法进行级联查询,方法参数为 column 的值,指向一个级联查询方法。

第 5 章　MyBatis 的关联关系

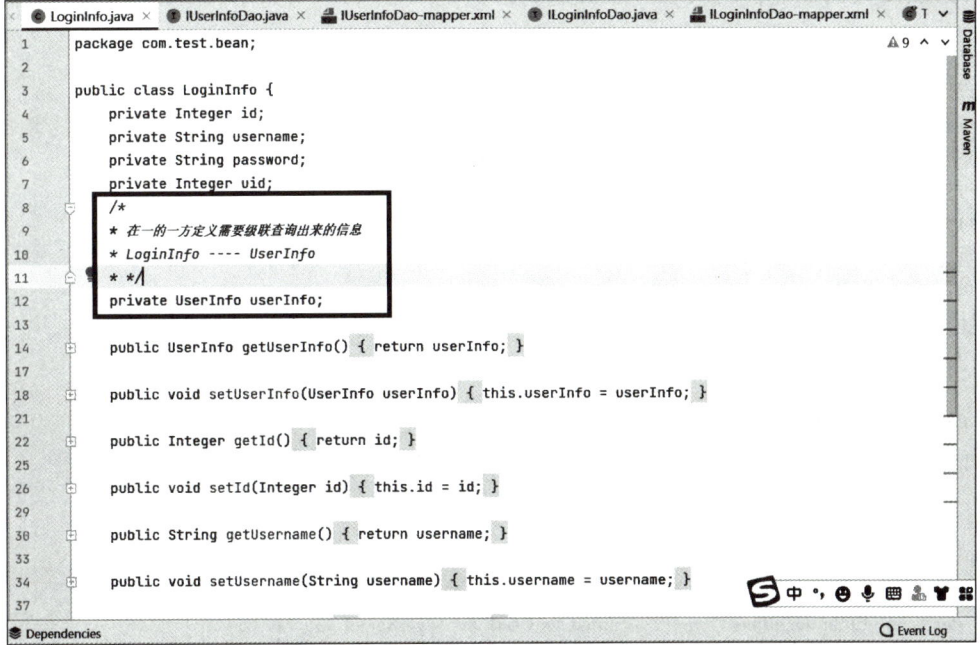

图 5-3　LoginInfo 类

(9) 将映射文件注册到 MyBatis 的配置文件中，如图 5-4 所示。

```
<mapper resource="ILoginInfoDao-mapper.xml"></mapper>
<mapper resource="IUserInfoDao-mapper.xml"></mapper>
```

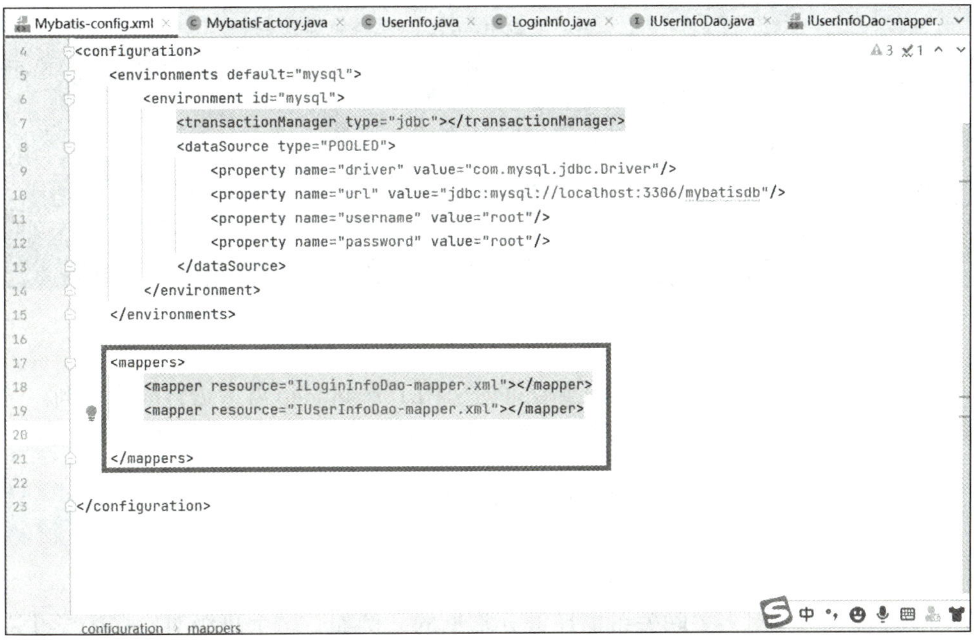

图 5-4　MyBatis 的配置文件

— 45 —

（10）测试案例如下，测试结果如图 5-5 所示。

```java
package com.test.test;
import com.test.bean.LoginInfo;
import com.test.dao.ILoginInfoDao;
import com.test.factory.MyBatisFactory;
import org.apache.ibatis.session.SqlSession;
import org.junit.Before;
import org.junit.Test;
public class TestDao{
    ILoginInfoDao dao;
    @Before
    public void before(){
        SqlSession session = MyBatisFactory.openSession();
        dao = session.getMapper(ILoginInfoDao.class);
    }
    @Test
    public void testLogin(){
        LoginInfo loginInfo = dao.login("admin","admin");
        System.out.println(loginInfo+"---"+loginInfo.getUserInfo().getName());
    }
}
```

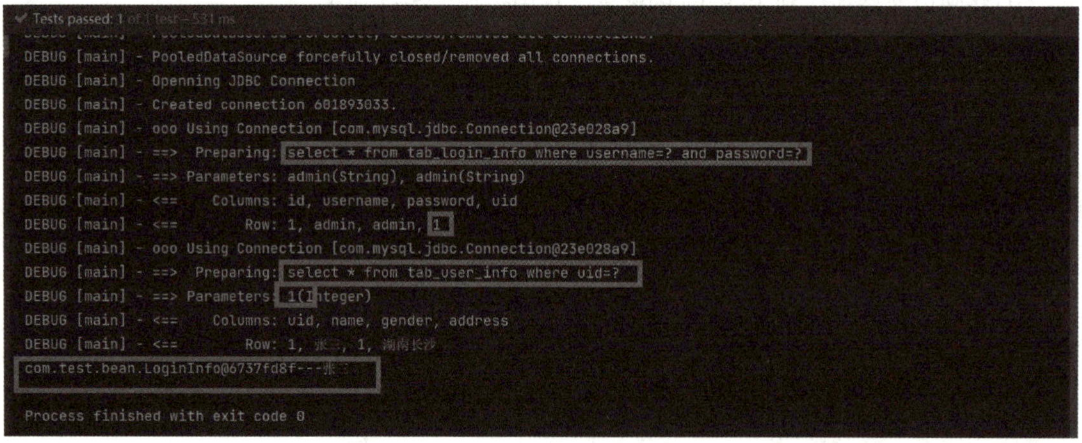

图 5-5　测试案例及结果

以上为一对一的案例代码。后期只要查询时引用了 resultMap 就会进行级联查询。

5.3　一对多

在开发过程中，一对多关联关系也是十分常见的。例如，一个班级对应多个学生，或者一个公司对应多个员工等的案例。接下来我们将以一个班级对应多个学生为例，来为大家讲解在 MyBatis 中如何实现一对多的级联关系查询。

在 MyBatis 中创建班级表(tab_classes)和学生表(tab_student)。数据如图 5-6 所示。

```
USE mybatisdb;
CREATE TABLE tab_classes(
    cid INT PRIMARY KEY AUTO_INCREMENT,
     cname VARCHAR(10) NOT NULL
);
CREATE TABLE tab_student(
    sid  INT PRIMARY KEY AUTO_INCREMENT,
    sname VARCHAR(8) NOT NULL,
    gender INT DEFAULT 1,
    cid INT NOT NULL,
    address VARCHAR(100)
);
```

cid	cname
1	软件1班
2	软件2班

sid	sname	gender	cid	address
1	张三	1	1	湖南长沙
2	李四	1	2	湖南湘潭
3	李丽	0	1	湖南长沙
4	赵敏	0	2	湖南XX

图 5-6　班级表和学生表

(1)在一对一项目中创建班级和学生的实体类。

```
package com.test2.bean;
public class Student {
private Integer sid;
private String sname;
private Integer gender;
private Integer cid;
private String address;
public Integer getSid() {
return sid;
    }
public void setSid(Integer sid) {
this.sid = sid;
    }

public String getSname() {
return sname;
    }
```

```java
    public void setSname(String sname) {
        this.sname = sname;
    }
    public Integer getGender() {
        return gender;
    }
    public void setGender(Integer gender) {
        this.gender = gender;
    }
    public Integer getCid() {
        return cid;
    }
    public void setCid(Integer cid) {
        this.cid = cid;
    }
    public String getAddress() {
        return address;
    }
    public void setAddress(String address) {
        this.address = address;
    }
}

package com.test2.bean;
import java.util.List;

public class Classes {
    private Integer cid;
    private String cname;
    /*
     * 一个班级包含多个学生,所以在班级中定义学生集合,级联查询出来的学生数据存放到该集合
     * 一个班级    对应    多个学生
     */
    private List<Student> students;
    public List<Student> getStudents() {
        return students;
    }

    public void setStudents(List<Student> students) {
        this.students = students;
    }
    public Integer getCid() {
        return cid;
    }
    public void setCid(Integer cid) {
        this.cid = cid;
    }
    public String getCname() {
```

```
        return cname;
    }
    public void setCname(String cname){
        this.cname = cname;
    }
}
```

一个班级包含多个学生，所以在班级中定义学生集合，将级联查询出来的学生数据存放到该集合。

（2）创建学生的数据访问层接口（IStudentDao）以及接口所对应的映射文件（IStudentDao-mapper.xml）。

```
package com.test2.dao;
import com.test2.bean.Student;
import java.util.List;
public interface IStudentDao{
    /**
     * 根据cid查询学生
     * @param cid
     * @return
     */
    public List<Student> queryStudentByCid(Integer cid);
}
```

映射文件如下所示。

```
<!DOCTYPE mapper PUBLIC
 "-//mybatis.org//DTD Mapper 3.0//EN"
 "http://mybatis.org/dtd/mybatis-3-mapper.dtd">
<mapper namespace="com.test2.dao.IStudentDao">
    <select id="queryStudentByCid"
        resultType="com.test2.bean.Student"
        parameterType="int">
        select * from tab_student where cid=#{cid}
    </select>
</mapper>
```

在查询班级时，需要根据班级id级联查询出所在班级信息，所以仅需创建queryStudentByCid方法即可。

（3）创建班级的数据访问层接口（IClassesDao）以及接口所对应的映射文件（IClassesDao-mapper.xml），如下所示。

```
package com.test2.dao;
import com.test2.bean.Classes;
public interface IClassesDao{
    /**
     * 根据cid查询班级信息
```

```
* @param cid
* @return
*/
public Classes queryClassesByCid(Integer cid);
}
```

```xml
<!DOCTYPE mapper PUBLIC
"-//mybatis.org//DTD Mapper 3.0//EN"
"http://mybatis.org/dtd/mybatis-3-mapper.dtd">
<mapper namespace="com.test2.dao.IClassesDao">
<!--
在查询班级时需要级联查询出所在班级的所有学生,查询出来的数据无法自动封装到classes实体类,所以需要定义resultMap手动封装
-->
<resultMap id="rm_classes" type="com.test2.bean.Classes">
<result property="cid" column="cid"></result>
<result property="cname" column="cname"></result>
<collection property="students" column="cid"
select="com.test2.dao.IStudentDao.queryStudentByCid">
</collection>
</resultMap>
<!--
在调用queryClassesByCid方法查询时需要级联查询出所在班级的学生信息,所以需要使用resultMap封装数据
-->
<select id="queryClassesByCid" resultMap="rm_classes" parameterType="int">
    select * from tab_classes where cid=#{cid}
</select>
</mapper>
```

因在查询班级时需要级联查询出所在班级的所有学生信息,并且将级联查询出来的信息封装到Classes中,但直接使用resultType无法满足,所以需要使用resultMap手动封装。

定义resultMap时需要在resultMap中用到<collection>标签进行级联查询操作。<collection>标签中属性功能如下。

property:级联查询出来的内容,指向实体类中属性。

column:级联查询时的条件,指向字段名。

select:使用数据访问层中的哪个方法进行级联查询,方法参数为column的值,指向一个级联查询方法。

(4) 将映射文件注册到MyBatis的配置文件中,如图5-7所示。

```xml
<mapper resource="IStudentDao-mapper.xml"></mapper>
<mapper resource="IClassesDao-mapper.xml"></mapper>
```

第 5 章　MyBatis 的关联关系

```
<!DOCTYPE configuration PUBLIC
    "-//mybatis.org//DTD Config 3.0//EN"
    "http://mybatis.org/dtd/mybatis-3-config.dtd">
<configuration>
    <environments default="mysql">
        <environment id="mysql">
            <transactionManager type="jdbc"></transactionManager>
            <dataSource type="POOLED">
                <property name="driver" value="com.mysql.jdbc.Driver"/>
                <property name="url" value="jdbc:mysql://localhost:3306/mybatisdb"/>
                <property name="username" value="root"/>
                <property name="password" value="root"/>
            </dataSource>
        </environment>
    </environments>

    <mappers>
        <mapper resource="ILoginInfoDao-mapper.xml"></mapper>
        <mapper resource="IUserInfoDao-mapper.xml"></mapper>
        <mapper resource="IStudentDao-mapper.xml"></mapper>
        <mapper resource="IClassesDao-mapper.xml"></mapper>
    </mappers>
```

图 5-7　MyBatis 的配置文件

（5）测试代码如下，测试结果如图 5-8 所示。

```
package com.test2.test;
import com.test.bean.LoginInfo;
import com.test.dao.ILoginInfoDao;
import com.test.factory.MyBatisFactory;
import com.test2.bean.Classes;
import com.test2.dao.IClassesDao;
import org.apache.ibatis.session.SqlSession;
import org.junit.Before;
import org.junit.Test;
public class TestDao{
    IClassesDao dao;
    @Before
    public void before(){
        SqlSession session = MyBatisFactory.openSession();
        dao = session.getMapper(IClassesDao.class);
    }
    @Test
    public void testQueryClassesByCid(){
        Classes cls = dao.queryClassesByCid(1);
        System.out.println(cls.getCid()+"--"+cls.getCname());
        cls.getStudents().forEach(s->System.out.println(s.getSname()));
    }
}
```

图 5-8 测试结果

以上为一对多代码案例,在查询班级时级联查询出所在班级的学生信息,主要使用 resultMap 中<collection>标签完成。

本章小结

> 本章首先对开发中涉及的数据表之间以及对象之间的关联关系作了简要介绍,并由此引出了 MyBatis 框架中对关联关系的处理;然后通过案例对 MyBatis 框架处理实体对象之间的关联关系进行了详细讲解。通过本章的学习,读者可以了解数据表以及对象中所涉及的关联关系,并能够使用 MyBatis 框架对关联关系的查询进行处理。MyBatis 中的关联查询操作在实际开发中非常普遍,熟练掌握这种关联查询方式有助于提高项目的开发效率,因此读者一定要多加练习。

课后习题

1. 请简述不同对象之间的三种关联关系。
2. 请简述 MyBatis 关联查询映射的两种处理方式。

第 6 章 Spring 的基本应用

学习目标

[本章知识点]
　　Spring 框架体系结构
　　Spring 框架布局
[思政目标]
　　提高学生自我学习和持续学习的意识和能力，使学生理解全局观的重要性，培养学生大局观意识。

6.1 认识 Spring 框架

　　曾经使用过 EJB 开发 JavaEE 应用的读者一定知道，EJB 开始的学习和应用非常的艰苦，很多东西都不能一下子就理解。EJB 要严格地实现各种不同类型的接口，类似的或者重复的代码大量存在，配置也复杂和单调，使用 JND1 进行对象查找的代码也是单调而枯燥的。虽然有一些开发工作随着 XDoclet 的出现，工作难度有所缓解，但是学习 EJB 高昂的代价、极低的开发效率、极高的资源消耗，都造成了 EJB 的使用困难，而 Spring 的出现就是为了解决这些类似的问题。

　　Spring 的一个最大作用就是使 JavaEE 开发更加容易。Spring 致力于提供一个以统一的、高效的方式构造整个应用，并且可以将单层框架以最佳的组合融合在一起建立连贯的体系。可以说 Spring 是一个提供了更完善开发环境的框架，可以为 FOO 对象提供企业级的服务。

6.2　Spring 框架优点

为使 JavaEE 的开发更容易、更简单，Spring 一直贯彻并遵守"好的设计优于具体实现，代码更易于测试"这一理念，并带给我们一个易于开发、便于测试而又功能齐全的开发框架。概括起来，Spring 给我们带来以下几方面的好处。

(1)方便解耦，简化开发。通过 Spring 提供的 IoC 容器，可以将对象之间的依赖关系交由 Spring 进行控制，避免硬编码所造成的程序耦合。有了 Spring，用户就不必再为单实例模式类、属性文件解析等这些很底层的需求编写代码，而可以更加专注于上层的应用。

(2)AOP 编程的支持。通过 Spring 提供的 AOP 功能，用户可以方便地进行面向切面的编程，许多不容易用传统面向对象编程实现的功能都可以通过 AOP 轻松实现。

(3)声明式事务的支持。在 Spring 中，用户可以从单调乏味的事务管理代码中解脱出来，通过声明式事务灵活进行事务管理，提高开发效率和质量。

(4)方便程序的测试。用户可以用非容器依赖的编程方式进行几乎所有的测试工作，在 Spring 中，测试不再是昂贵的操作，而是随手可做的事情。

(5)方便集成各种优秀的框架。Spring 不排斥各种优秀的开源框架，相反，Spring 可以降低各种框架的使用难度。Sping 提供了对各种优秀框架(如 Structs、Hibernate、Hessian、Quartz 等)的直接支持。

(6)降低 JavaEE API 的使用难度。Spring 为很多难用的 JavaEE API(如 JDBC、Java Mail、远程调用等)提供了一个薄薄的封装层，通过 Spring 的简易封装，可以大大降低这些 JavaEE API 的使用难度。

(7)Java 源码式经典的学习范例。Spring 的源码设计精妙、结构清晰、匠心独具，处处体现着大师对设计模式的灵活运用以及对 Java 技术的高深造诣。Spring 框架源码无疑是 Java 技术的最佳实践范例。如果想在短时间内迅速提高自己的 Java 技术水平和应用开发水平，那么学习和研究 Spring 源码就可以让你获得意想不到的效果。

6.3　Spring 框架体系结构

Spring 框架是一个分层架构，它包含一系列的功能要素并被分为大约 20 个模块。这些模块分为 Core Container、Data Access/Integration、Web、AOP(Aspect Oriented Programming)、Aspects、Instrumentation 和 Test，如图 6-1 所示。

第 6 章　Spring 的基本应用

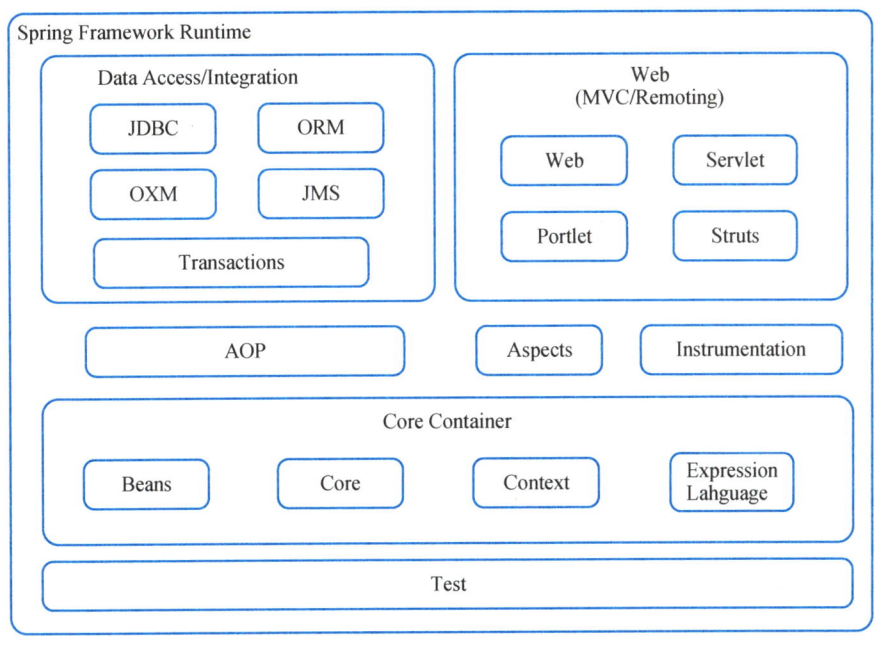

图 6-1　Spring 框架体系结构图

6.3.1　核心容器（Core Container）

Core 和 Beans 模块提供了 Spring 最基础的功能，提供 IoC 和依赖注入特性。这里的基础概念是 BeanFactory，它提供对 Factory 模式的经典实现来消除对程序性单例模式的需要，并真正地允许你从程序逻辑中分离出依赖关系和配置。

Context 模块基于 Core 和 Bean 来构建，它提供了用一种框架风格的方式来访问对象，有些像 JNDI 注册表。Context 封装包继承了 Beans 包的功能，还增加了国际化（I18N）、事件传播、资源装载，以及透明创建上下文等功能。

Expression Language，表达式语言模块，提供了在运行期间查询和操作对象的强大能力。支持访问和修改属性值，支持方法调用，支持访问及修改数组、容器和索引器，支持命名变量，支持算数和逻辑运算，支持从 spring 容器获取 Bean，还支持列表投影、选择和一般的列表聚合等。

6.3.2　数据访问/集成部分（Data Access/Integration）

JDBC 模块，提供对 JDBC 的抽象，它可消除冗长的 JDBC 编码，解析数据库厂商特有的错误代码。

ORM 模块，提供了常用的"对象/关系"映射 API 的集成层。其中包括 JPA、JDO、Hibernate 和 iBatis。利用 ORM 封装包，可以混合使用所有 Spring 提供的特性进行"对象/关系"映射,如简单声明式事务管理。

OXM 模块,提供一个支持 Object 和 XML 进行映射的抽象层。其中包括 JAXB、Castor、XMLBeans、JiBX 和 XStream。

JMS 模块,提供一套"消息生产者、消费者"模板用于更加简单地使用 JMS,JMS 用于在两个应用程序之间或分布式系统中发送消息,进行异步通信。

Transaction 模块,支持程序的简单声明式事务管理,只要是 Spring 管理对象都能得到 Spring 管理事务的好处。

6.3.3 Web

web-socket 模块:websocket protocol 是 HTML5 一种新的协议,它实现了浏览器与服务器全双工通信,Spring 支持 websocket 通信。

web 模块:提供了基础的 Web 功能,例如多文件上传、集成 IoC 容器、远程过程访问以及对 webservice 支持,并提供一个 RestTemplate 类来提供方便的 Restful services 访问。

web-servlet 模块:提供了 Web 应用的 model-view-controller(MVC)实现。Spring MVC 框架提供了基于注解的请求资源注入,更简单的数据绑定、数据验证以及一套非常易用的 JSP 标签,能够完全无缝与 Spring 其他技术协作。

web-portlet 模块:提供了在 portlet 环境下的 MVC 实现。

6.3.4 AOP

AOP 模块:提供了符合 AOP 联盟规范的面向切面的编程实现,让用户可以定义方法拦截器和切入点,从逻辑上讲,可以减弱代码的功能耦合。而且,利用源码级的元数据功能,还可以将各种行为信息合并到用户的代码中。

Aspects 模块:提供了对 AspectJ 的集成。

Instrumentation 模块:提供一些类级的工具支持和 ClassLoader 级的实现,可以在一些特定的应用服务器中使用。

6.3.5 Test

Test 模块支持使用 JunitNG 对 Spring 组件进行测试。

第 6 章　Spring 的基本应用

6.4　Spring 项目布局

6.4.1　创建 Maven 工程

新建 Maven 工程，不使用 Maven 模板，如图 6-2 所示。

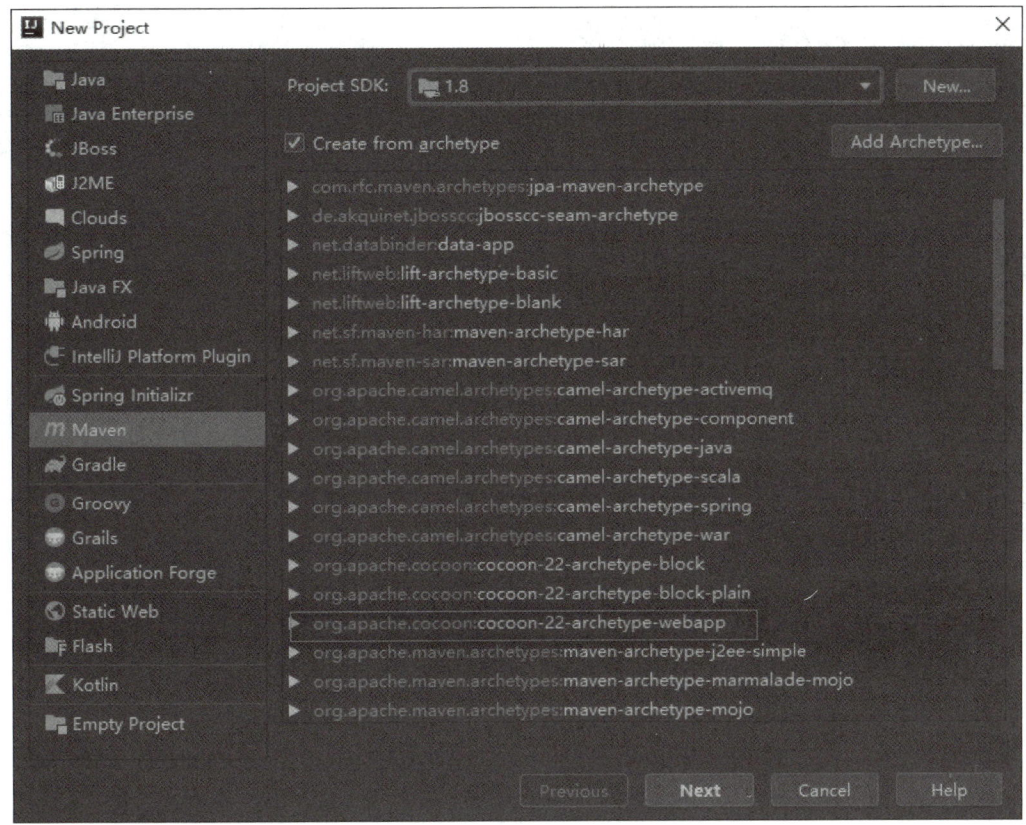

图 6-2　新建 Maven 项目

点击 Next 后，填写相关的 GroupId 和 ArtifactId 后，再一次点击 Finish，如图 6-3 所示。

— 57 —

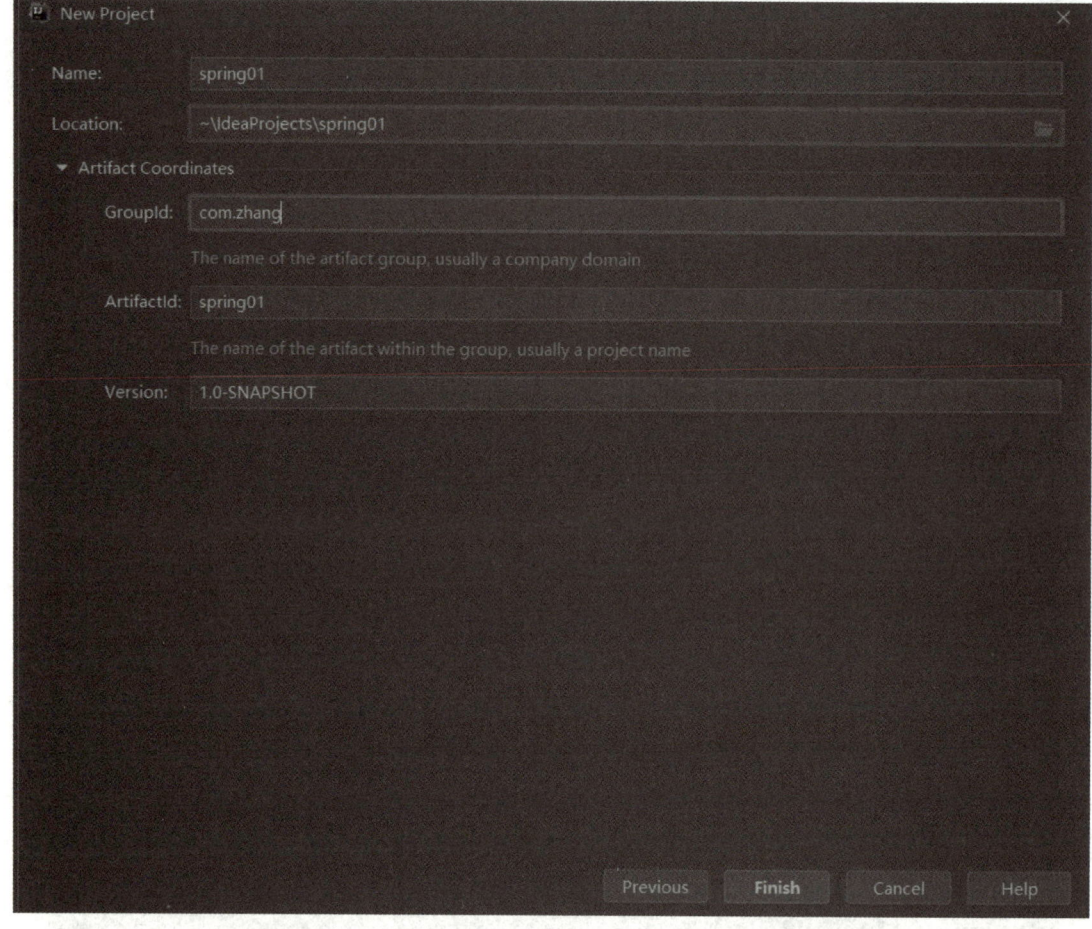

图 6-3 GroupId 和 ArtifactId

在设置 User setting file 中，因为生成 Maven 时默认是没有 settings.xml 文件的，可以在 IDEA 的安装目录里复制一份过来，在 settings.xml 文件里添加如下镜像。

```
<mirror>
<id>alimaven</id>
<mirrorOf>central</mirrorOf>
<name>aliyun maven</name>
<url>http://maven.aliyun.com/nexus/content/repositories/central/</url>
</mirror>
```

6.4.2 修改 pom.xml 文件，添加依赖

刚开始时项目目录如图 6-4 所示。

第 6 章　Spring 的基本应用

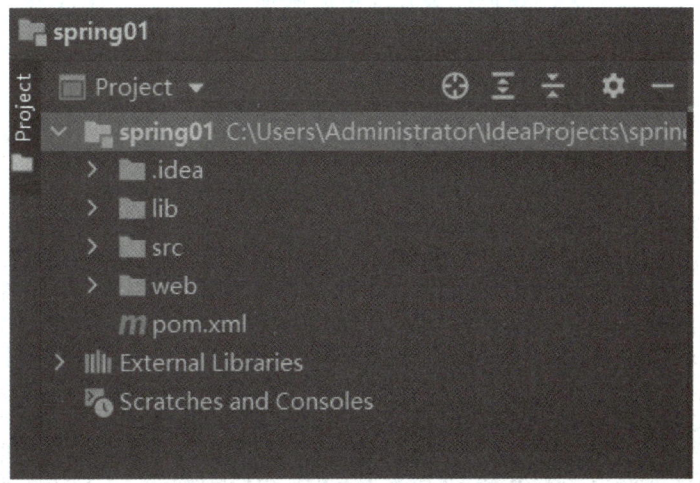

图 6-4　项目目录

打开 pom.xml 文件添加相应的依赖，本文的依赖如下所示。

```
<dependencies>
<dependency>
<groupId>org.springframework</groupId>
<artifactId>spring-context</artifactId>
<version>5.1.7.RELEASE</version>
</dependency>
</dependencies>
```

6.4.3　加载项目

配置完 pom.xml 之后，在工程 spring01 上右键选择 Maven，点击 reload project，如图 6-5 所示。

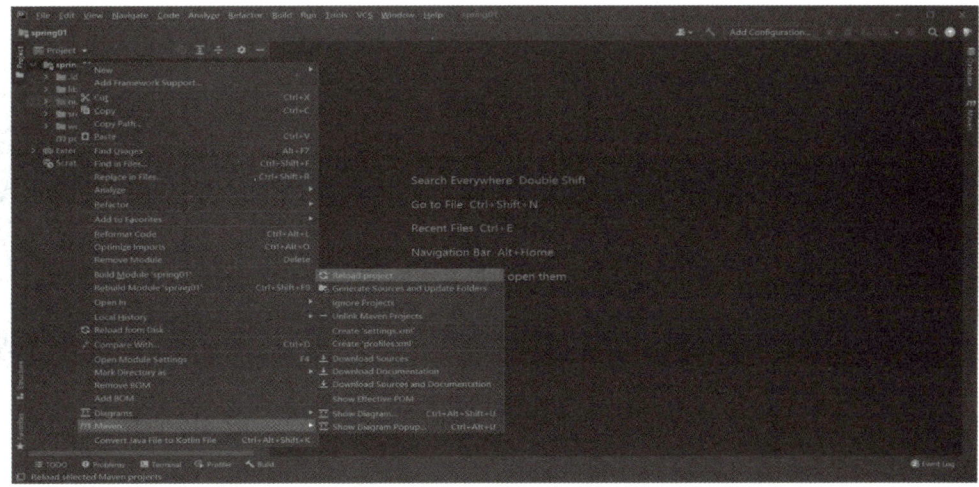

图 6-5　加载项目

加载完之后，Spring 容器会根据 pom.xml 里面添加的 spring-context 核心依赖，找到相关 jar 包，如图 6-6 所示。

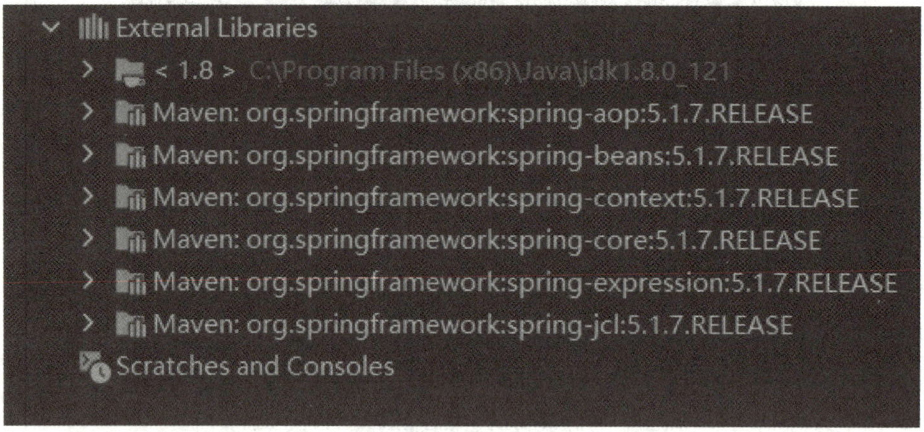

图 6-6　Spring jar 包

完成后，在 resources 下会自动生成 spring.xml 文件，如图 6-7 所示。

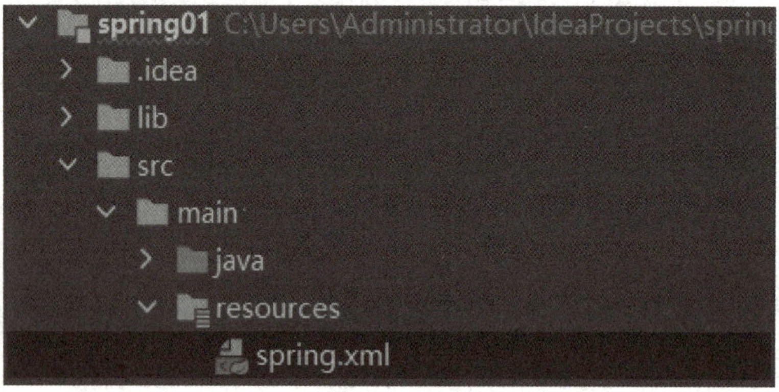

图 6-7　自动生成 spring.xml 文件

打开 spring.xml 可以看到 Beans 中的内容也相应地出现，如图 6-8 所示。

图 6-8　spring.xml

6.5 Spring 核心容器

Spring 是作为一个容器存在的，应用中的所有组件都处于 Spring 的管理之下，都被 Spring 以 Bean 的方式管理。

Spring 负责创建 Bean 的实例，并管理其生命周期。Spring 框架的主要功能是通过其核心容器来实现的。Spring 有两个核心接口，分别为 BeanFactory 和 ApplicationContext。其中 ApplicationContext 是 BeanFactory 的子接口，它们都可代表 Spring 容器。也就是说 Spring 框架提供了两种核心容器，即 BeanFactory 和 ApplicationContext。

6.5.1 BeanFactory

BeanFactory 由 org.springframework.beans.facytory.BeanFactory 接口定义，是基础类型的 IoC 容器(IoC 控制反转)，它提供了完整的 IoC 服务支持。简单来说，BeanFactory 就是一个管理 Bean 的工厂，它主要负责初始化各种 Bean，并调用它们的生命周期方法。

BeanFactory 接口提供了几个实现类，其中最常用的是 org.springframework.beans.factory.xml.XmlBeanFactory，该类会根据 XML 配置文件中的定义来装配 Bean。

创建 BeanFactory 实例时，需要提供 Spring 所管理容器的详细配置信息，这些信息通常采用 XML 文件形式来管理，其加载配置信息的语法如下。

BeanFactorybeanFactory =new XmlBeanFactory(newFileSystemResource("D:/applicationContext.xml"));

D:/applicationContext.xml 是 XML 配置文件的位置。这种加载方式在实际开发中并不多见，了解即可。

6.5.2 ApplicationContext

ApplicationContext 是 BeanFactory 的子接口，也被称为"应用上下文"，是另一种常用的 Spring 核心容器。它由 org.springframework.context.ApplicationContext 接口定义，不仅包含了 BeanFactory 的所有功能，还添加了对国际化、资源访问、事件传播等方面的支持。

ApplicationContext 和 BeanFacotry 两者都是用于加载 Bean 的，但是相比之下 ApplicationContext 提供了更多的扩展功能，简单一点说，ApplicationContext 包含 BeanFactory 的所有功能。通常建议 ApplicationContext 比 BeanFactory 优先使用，除非在一些限制的场合，比如字节长度对内存有很大的影响时(Applet)。绝大多数企业应用和系统，都是使用 ApplicationContext 的。

创建 ApplicationContext 接口实例，通常采用两种方法，具体如下。

1. 通过 ClassPathXmlApplicationContext 创建

ClassPathXmlApplicationContext 会从类路径 ClassPath 中寻找指定的 XML 配置文件，找到并装载完成 ApplicationContext 的实例化工作，其使用语法如下。

```
//初始化 Spring 容器,加载配置文件
    ApplicationContextapplicationContext =
new ClassPathXmlApplicationContext(String configLocation);
//初始化 Spring 容器,加载配置文件
    ApplicationContextapplicationContext =
new ClassPathXmlApplicationContext("applicationContext.xml");
```

上述代码中，configLocation 参数用于指定 Spring 配置文件的名称和位置。如果其值为"applicationContext.xml"，则 Spring 会去类路径中查找名称为 applicationContext.xml 的配置文件。

2. 通过 FileSystemXmlApplicationContext 创建

FileSystemXmlApplicationContext 会从指定的文件系统路径(绝对路径)中寻找指定的 XML 配置文件，找到并装载完成 ApplicationContext 的实例化工作，其使用语法如下。

```
//初始化 Spring 容器,加载配置文件
    ApplicationContextapplicationContext =
new FileSystemXmlApplicationContext(String configLocation);
//初始化 spring 容器,加载配置文件
    ApplicationContextapplicationContext =
new FileSystemXmlApplicationContext("applicationContext.xml");
```

与 ClassPathXmlApplicationContext 有所不同的是，在读取 Spring 的配置文件时，FileSystemXmlApplicationContext 不再从类路径中读取配置文件，而是通过参数指定配置文件的位置，例如"D:/workspaces/applicationContext.xml"。如果在参数中写的不是绝对路径，那么方法调用的时候，会默认用绝对路径来找。

这种采用绝对路径的方式，会导致程序的灵活性变差，所以这个方法一般不推荐使用。

在使用 Spring 框架时，可以通过实例化其中任何一个类来创建 ApplicationContext 容器。

通常在 Java 项目中，会采用通过 ClassPathXmlApplicationContext 类来实例化 ApplicationContext 容器的方式，而在 Web 项目中，ApplicationContext 容器的实例化工作会交由 Web 服务器来完成。Web 服务器实例化 ApplicationContext 容器时，通常会使用基于 ContextLoaderListener 实现的方式，此种方式只需要在 web.xml 中添加如下代码。

```xml
<context-param>
<param-name>contextConfigLocation</param-name>
<param-value>
classpath:spring/applicationContext.xml
</param-value>
```

```
</context-param>
<listener>
<listener-class>
org.springframework.web.context.ContextLoaderListener
</listener-class>
</listener>
```

创建 Spring 容器后，就可以获取 Spring 容器中的 Bean。Spring 获取 Bean 的实例通常采用以下两种方法。

• Object getBean(String name)：根据容器中 Bean 的 id 或 name 来获取指定的 Bean，获取之后需要进行强制类型转换。

• T getBean(Class requiredType)：根据类的类型来获取 Bean 的实例。由于此方法为泛型方法，因此在获取 Bean 之后不需要进行强制类型转换。

BeanFactory 和 ApplicationContext 两种容器都是通过 XML 配置文件加载 Bean 的。二者的主要区别在于，如果 Bean 的某一个属性没有注入，使用 BeanFacotry 加载后，在第一次调用 getBean()方法时会抛出异常；而 ApplicationContext 则在初始化时自检，这样有利于检查所依赖属性是否注入。因此，在实际开发中，通常都优先选择使用 ApplicationContext，而只有在系统资源较少时，才考虑使用 BeanFactory。

6.6 Spring 入门程序

在 6.4 节已经创建好项目，Spring 所需要的 jar 也已经引入过来，Spring.xml 已经生成。下面通过一个简单的入门程序来演示 Spring 框架的使用。

（1）在项目的 main 文件夹下找到 java 文件，在下面创建包 com.yyzy.ioc，如图 6-9 所示。

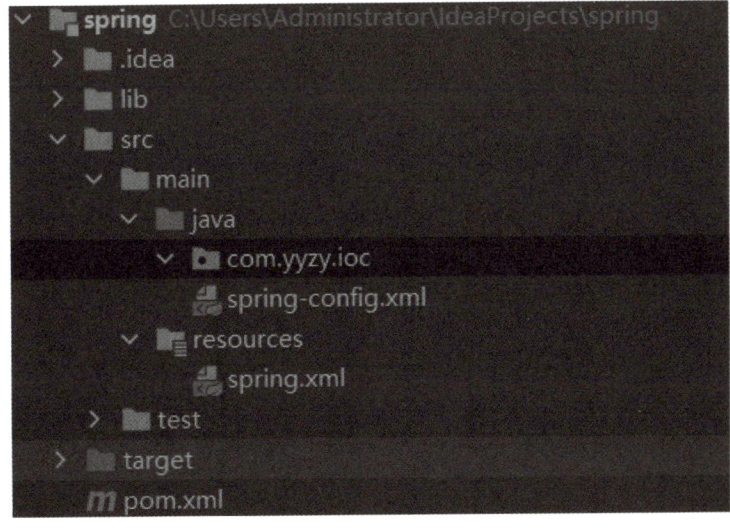

图 6-9　创建包

(2) 在该包下创建类 Dog，然后在类中定义一个 show( ) 方法。

```
public class Dog {
    public void show( ) {
System.out.println("我有一个小白狗");
    }
}
```

(3) 在 resources 文件夹下的 spring.xml 配置文件中，创建一个 id 为 dog 的 Bean。

```
<?xml version="1.0" encoding="UTF-8"?>
<beans xmlns="http://www.springframework.org/schema/beans"
xmlns:xsi="http://www.w3.org/2001/XMLSchema-instance"
xsi:schemaLocation="http://www.springframework.org/schema/beans
        http://www.springframework.org/schema/beans/spring-beans.xsd">
<bean id="dog" class="com.yyzy.ioc.Dog"></bean>
</beans>
```

(4) 在 com.yyzy.ioc 包下创建测试类 TestDog，并在类中编写 main 方法。在 main 方法中，需要初始化 Spring 容器，并加载配置文件，然后通过 Spring 容器获取 dog 实例，最后调用实例中的 show( ) 方法。

```
public class TestDog {
    public static void main(String[ ] args) {
ApplicationContext context = new ClassPathXmlApplicationContext("spring.xml");
        Dog dog = (Dog)context.getBean("dog");
dog.show( );
    }
}
```

本章小结

本章主要讲解了 Spring 框架入门的一些基础知识。首先讲解了 Spring 框架的概念、作用、优点、体系结构以及如何下载和下载后的目录结构，然后介绍了 Spring 的两种核心容器，最后通过一个入门程序来讲解如何使用 Spring 框架。

课后习题

1. 请简述 Spring 框架的优点。
2. 请简述 Spring 的核心容器有哪些。

第7章 Spring 框架之 IoC

 学习目标

[本章知识点]
  Spring 框架对于 Bean 的基本操作
  依赖注入
  Spring 注解操作 Bean

[思政目标]
  使学生理解全局观的重要性，培养大局意识。
  使学生通过对 Spring 框架的学习，加深对专业知识技能学习的认可度与专注度。

7.1 Spring 操作 Bean

要使应用程序中的 Spring 容器成功启动，需要同时具备以下 3 方面的条件：

（1）Spring 框架的类包都已经正确放到应用程序的类路径下。

（2）应用程序为 Spring 提供完备的 Bean 配置信息。

（3）Bean 的类都已经正确放到应用程序的类路径下。

Spring 启动时读取应用程序提供的 Bean 配置信息，并在 Spring 容器中生成一张相应的 Bean 配置注册表，然后根据这张注册表实例化 Bean，装配好 Bean 之间的依赖关系，为上层应用提供准备就绪的运行环境。

Bean 配置信息由 4 部分组成：

（1）Bean 的实现类。

（2）Bean 的属性信息。如数据源的连接数、用户名和密码等。

（3）Bean 的依赖关系。Spring 根据依赖关系配置完成 Bean 之间的装配。

(4) Bean 的行为配置。如生命周期范围及生命周期各过程的回调函数等。

7.2 创建 Bean

一般情况下，Spring IoC 容器中的一个 Bean 即对应配置文件中的一个 <bean>。Spring 框架使用配置文件方式创建 Bean 的方法有三种：构造器方法、静态工厂方法、实例工厂方法。

下面通过一个案例来演示 Spring 容器是如何实例化 Bean 的。

在 IDEA 中，创建一个名为 chapter07 的 Web 项目，在该项目中添加依赖 jar 包。

在 chapter07 项目的 src 目录下，创建一个 com.yyzy.bean.po 包，在该包中创建 Cat 类。定义属性、set 和 get 方法、有参的构造方法和无参的构造方法。

7.2.1 构造器方法

构造器实例化是指 Spring 容器通过 Bean 对应类中默认的无参构造方法来实例化 Bean。

```
<!--Cat cat=new Cat();-->
<bean id="cat" class="com.yyzy.bean.po.Cat"></bean>
```

在 com.yyzy.bean.po 包中，创建测试类 TestCat，来测试构造器是否能实例化 Bean。

```
public static void main(String[] args) {
//定义配置文件路径
String xmlPath = "com/itheima/instance/constructor/beansl.xml";
// ApplicationContext 在加载配置文件时,对 Bean 进行实例化
ApplicationContext applicationContext =
new ClassPathXmlApplicationContext(xmlPath);
Cat cat = (Cat) applicationContext.getBean("cat");
System.out.println(cat);
}
```

首先定义配置文件的路径，然后 Spring 容器的 ApplicationContext 会加载配置文件。在加载时，Spring 容器会通过 id 为 beanl 的实现类 Beanl 中默认的无参构造方法对 Bean 进行实例化。

7.2.2 静态工厂方法

使用静态工厂方法是实例化 Bean 的另一种方式。该方式要求开发者以创建一个静态工厂的方法来创建 Bean 的实例，其 Bean 配置中的 class 属性所指定的不再是 Bean 实例的实现类，而是静态工厂类，同时还需要使用 factory-method 属性来指定所创建的静态工厂方法。

在 chapter07 项目的 src 目录下，创建一个 com.yyzy.bean.factory 包，在该包中创建一个 Cat2 类，该类与 Cat 一样，不需添加任何方法。

在 com.yyzy.bean.factory 包中，创建一个 CatFactory 类，并在类中创建一个静态方法 getStaticCat( ) 来返回 Cat 实例。

```
public class CatFactory{
//使用自己的工厂创建 Cat 实例
public static  CatgetStaticCat( ){
Cat    cat=new Cat( );
return cat;
}
}
```

配置文件如下所示。

```
<bean id="cat2"
class="com.yyzy.bean.factory.CatFactory"
factory-method="getStaticCat" />
</beans>
```

在上述配置文件中，首先通过<bean>元素的 id 属性定义了一个名称为"cat2"的 Bean，然后由于使用的是静态工厂方法，所以需要通过 class 属性指定其对应的工厂实现类为 CatFactory。由于以这种方式配置 Bean 后，Spring 容器不知道哪个是所需要的工厂方法，所以需要增加 factory-method 属性来告诉 Spring 容器，其方法名称为"getStaticCat"。

7.2.3 实例工厂方法

还有一种实例化 Bean 的方式就是采用实例工厂。此种方式的工厂类中，不再使用静态方法创建 Bean 实例，而是采用直接创建 Bean 实例的方式。同时，在配置文件中，实例化 Bean 也不是通过 class 属性直接指向实例化类，而是通过 factory-bean 属性指向配置的实例工厂，然后使用 factory-method 属性确定使用工厂中的哪个方法。

在 chapter07 项目的 src 目录下，创建一个 com.yyzy.bean.factory 包，在该包中创建 Cat3 类，该类与 Cat 一样，不需添加任何方法。

在 com.yyzy.bean.factory 包中，创建工厂类 Cat3Factory，在类中使用默认无参构造方法输出 Cat3 工厂实例化中内容，并使用 createBean( ) 方法创建 Cat3 对象。

```
public class Cat3Factory{
public Cat3Factory( ){
System.out.println("Cat 工厂实例化中");
}
//创建 Cat3 实例的方法
public Cat3 createBean( ){
return new Cat3( );
}
}
```

创建 Spring 配置文件 beans3_xml，设置相关配置。

```xml
<!--配置工厂-->
<bean id="myBean3Factory"
class="com.yyzy.bean.factory.Cat3Factory"/>
<!--使用factory-bean属性指向配置的实例工厂,
使用factory-method属性确定使用工厂中的哪个方法-->
<bean id="cat3" factory-bean="myBean3Factory"
factory-method:"createBean"/>
</beans>
```

7.3　管理 Bean

通过 Spring 容器创建一个 Bean 的实例时，不仅可以完成 Bean 的实例化，还可以为 Bean 指定特定的作用域。本节将主要围绕 Bean 的作用域知识进行讲解。

7.3.1　作用域的种类

Spring 4.3 中为 Bean 的实例定义了 7 种作用域，通过属性 scope 设置作用域。这 7 种作用域及其说明如表 7-1 所示。

表 7-1　Bean 的作用域

作用域名称	说明
singleton（单例）	使用 singleton 定义的 Bean 在 Spring 容器中将只有一个实例，也就是说，无论有多少个 Bean 引用它，始终将指向同一个对象。这也是 Spring 容器默认的作用域
prototype（原型）	每次通过 Spring 容器获取 prototype 定义的 Bean 时，容器都将创建一个新的 Bean 实例
request	在一次 HTTP 请求中，容器会返回该 Bean 的同一个实例。对不同的 HTTP 请求则会产生一个新的 Bean，而且该 Bean 仅在当前 HTTP Request 内有效
session	在一次 HTTP Session 中，容器会返回该 Bean 的同一个实例。对不同的 HTTP 请求则会产生一个新的 Bean，而且该 Bean 仅在当前 HTTP Session 内有效
globalSession	在一个全局的 HTTP Session 中，容器会返回该 Bean 的同一个实例。仅在使用 portlet 上下文时有效
application	为每个 ServletContext 对象创建一个实例。仅在 Web 相关的 ApplicationContext 中生效
websocket	为每个 websocket 对象创建一个实例。仅在 Web 相关的 ApplicationContext 中生效

7.3.2　Bean 生命周期

Spring 容器可以管理 singleton 作用域的 Bean 的生命周期，在此作用域下，Spring 能够精确地知道该 Bean 何时被创建，何时初始化完成以及何时被销毁。对于 prototype 作用域的

Bean，Spring 只负责创建，当容器创建了 Bean 实例后，Bean 的实例就交给客户端代码来管理，Spring 容器将不再跟踪其生命周期。每次客户端请求 prototype 作用域的 Bean 时，Spring 容器都会创建一个新的实例，并且不会跟踪那些被配置成 prototype 作用域的 Bean 的生命周期。

Bean 生命周期：创建、初始化、服务、死亡。

Spring 在创建 Bean 实例后，会调用 Bean 的初始化方法。Spring 在容器关闭后，Bean 不被 Spring 容器调用，进入垃圾回收阶段，在容器关闭之前，会调用 Bean 的销毁方法。

```
public class Cat {
    public Cat() {
        System.out.println("create cat");
    }
    public void initCat() {
        System.out.println("init");
    }
    public void destroyCat() {
        System.out.println("destroy");
    }
    public void deadCat() {
        System.out.println("wowowo");
    }
}
public static void main(String[] args) {
    ClassPathXmlApplicationContext context = new ClassPathXmlApplicationCotext("cat");
    Cat cat = (Cat)context.getBean("cat");
    cat.deadCat();
}
```

在配置文件中设置，调用 deadCat() 后，生命周期执行顺序清晰可见。

```
<bean id="cat" class="com.yyzy.bean.po.Cat" init-method="initCat" destroy-method="destroyCat"></bean>
```

7.4 依赖注入

当某个 Java 对象（调用者）需要调用另一个 Java 对象（被调用者，即被依赖对象）时，在传统模式下，调用者通常会采用"new 被调用者"的代码方式来创建对象。这种方式会导致调用者与被调用者之间的耦合性增加，不利于后期项目的升级和维护。

从 Spring 容器的角度来看，Spring 容器负责将被依赖对象赋值给调用者的成员变量，这相当于为调用者注入了它依赖的实例，这就是 Spring 的依赖注入（Dependency Injection，DI）。

在使用 Spring 框架之后，对象的实例不再由调用者来创建，而是由 Spring 容器来创建，Spring 容器会负责控制程序之间的关系，而不是由调用者的程序代码直接控制。这样，控制权由应用代码转移到了 Spring 容器，控制权发生了反转，这就是 Spring 的控制反转（IoC）。

依赖注入的作用就是在使用 Spring 框架创建对象时，动态地将其所依赖的对象注入 Bean 组件中。属性注入方式如图 7-1 所示。

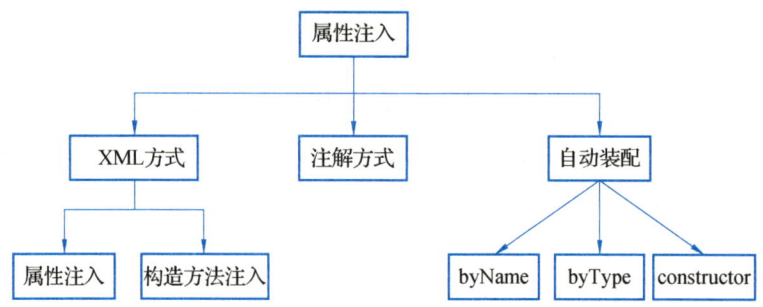

图 7-1　属性注入方法图

7.4.1　XML 注入方式

Spring 提供了两种基于 XML 的装配方式：设值注入（Setter Injection）和构造注入（Constructor Injection）。下面就来讲解如何在 XML 配置文件中使用这两种注入方式来实现基于 XML 的装配。

属性值注入即通过 setXxx() 方法注入 Bean 的属性值或依赖对象，要求 Bean 提供一个默认的构造方法，并为需要注入的属性提供对应的 setter 方法。Spring 先调用 Bean 的默认构造函数实例化 Bean 对象，然后通过反射的方式调用 setter 方法注入属性值。

1. 基本类型

在项目 chapter07 的 src 目录下，创建一个 com.yyzy.xml 包，在该包中创建 User 类，并在类中定义 username、sex、age 三个属性及其对应的 setter 方法。

```
public class User {
    private    String username;
    private String sex;
    private Integer age;
    public void setUsername(String username) {
this.username = username;
}
public void setAge(Integer age) {
this.age = age;
}
    public void setSex(String sex) {
this.sex = sex;
}
public String getUsername() {
```

```
        return username;
}
public String getSex( ) {
        return sex;
}
    public Integer getAge( ) {
        return age;
}
    public User( ) {
    }
}
```

setter 方式注入属性值，是在<bean>标签里面使用<property>标签，然后通过 name/value 属性注入值。

```
<bean id="user" class="com.yyzy.xml.user">
<property name="username" value="李宁"></property>
<property name="sex" value="男"></property>
<property name="age" value="23"></property>
</bean>
```

在 com.yyzy.xml 包中，创建测试类 TestUser。

```
public class TestUser {
    public static void main(String[] args) {
        //加载配置文件
ApplicationContextapplicationContext =
                new ClassPathXmlApplicationContext("bean.xml");
        //通过 Bean 生成 User 对象
User user = (User)applicationContext.getBean("user");
        //输出结果用户名
System.out.println(user.getUsername());
}
}
```

构造函数注入是除属性注入之外的另一种常用的注入方法。构造函数注入需在<bean>标签里面使用<constructor-arg>标签，<constructor-arg>标签可以通过属性 index/value 组合或者 name/value 组合来给属性注入值，此时调用对应的构造方法。其中，index 表示有参构造方法中参数的位置，从 0 开始算第一个参数。注入属性值时，需要有对应的参数构造方法，如果没有对应参数的构造方法，配置会报错。

通过构造方法同样为上面的 User 对象的三个属性赋值。User 对象需要增加构造方法。

```
public class User {
private    String username;
    private String sex;
    private    Integer age;
public User(String username, String sex, Integer age) {
```

```
        this.username = username;
        this.sex = sex;
        this.age = age;
    }
    public String getUsername() {
        return username;
    }
    public String getSex() {
        return sex;
    }
    public Integer getAge() {
        return age;
    }
}
```

在 resources 下创建 bean2.xml 文件。

```
<bean id="user2" class="com.yyzy.xml.User">
<constructor-arg index="0" value="tom"></constructor-arg>
<constructor-arg name="sex" value="女"></constructor-arg>
<constructor-arg index="2" value="23"></constructor-arg>
</bean>
```

2. 引用类型

基本类型主要指 String、int、float 等封装器类型，引用类型表示用户自定义的类对象，即引用 Spring 容器配置好的 Bean 组件，此时使用 ref 属性来代替 values 属性，表示引用。在前面的 User 类中增加一个属性 Cat。

```
public class User {
private  String username;
    private String sex;
   private  Integer age;
Cat  cat;
    public User() {
    }
public void setUsername(String username) {
this.username = username;
}
public void setSex(String sex) {
this.sex = sex;
}
public void setAge(Integer age) {
this.age = age;
}
public void setCat(Cat cat) {
        this.cat = cat;
    }
}
```

Bean.xml 文件如下所示。

```xml
<bean id="cat" class="com.yyzy.xml.Cat">
<property name="cname" value="小花猫"></property>
<property name="age" value="3"></property>
</bean>
<bean id="user" class="com.yyzy.xml.User">
<property name="username" value="李宁"></property>
<property name="sex" value="男"></property>
<property name="age" value="23"></property>
<property name="cat" ref="cat"></property>
</bean>
```

测试如下所示。

```java
public static void main(String[] args){
    //加载配置文件
    ApplicationContext applicationContext =
            new ClassPathXmlApplicationContext("bean.xml");
    //通过 bean 生成 User 对象
    User user=(User)applicationContext.getBean("user");
    //输出结果猫名
    System.out.println(user.getCat().getCname());
}
```

3. 集合、Map 等类型

集合类型注入指 Bean 组件的属性为 List、Map、数组等。

在 User 类中增加数组、集合和 Map 属性。

```java
public class User{
    private     String username;
    private String sex;
    private     Integer age;
int a[];
    List<Cat> listcat;
    Map<String,String> map1;
    Map<String,Cat> mapcat;
省略无参和有参构造方法,setter 和 getter 方法,toString 方法
}
```

Bean.xml 文件如下所示。

```xml
<bean id="cat" class="com.yyzy.xml.Cat">
<property name="cname" value="小花猫"></property>
<property name="age" value="3"></property>
</bean>
<bean id="cat2" class="com.yyzy.xml.Cat">
<property name="cname" value="小白猫"></property>
```

```xml
<property name="age" value="2"></property>
</bean>
<bean id="user" class="com.yyzy.xml.User">
<property name="username" value="李宁"></property>
<property name="sex" value="男"></property>
<property name="age" value="23"></property>
<!--为数组注入值-->
<property name="a">
<array>
<value>10</value>
<value>20</value>
<value>30</value>
<value>40</value>
</array>
</property>
<!--List 集合的数据类型为 String,需要使用 value-->
<property name="list">
<list>
<value>tom</value>
<value>rose</value>
<value>mike</value>
</list>
</property>
<!--List 集合的数据类型为自定义类型,需要使用 ref-->
<property name="listcat">
<list>
<ref bean="cat"></ref>
<ref bean="cat2"></ref>
</list>
</property>
<!--Map 集合属性赋值时,需要使用 entry-->
<property name="map1">
<map>
<entry key="SSM 框架" value="1本"/>
<entry key="javaWeb" value="5本"/>
<entry key="MySql 数据库" value="10本"/>
</map>
</property>
<!--Map 集合属性赋值时,需要使用 value-ref-->
<property name="mapcat">
<map>
<entry key="cat" value-ref="cat"/>
<entry key="cat2" value-ref="cat2"/>
</map>
</property>
</bean>
```

使用测试函数测试。

```
public static void main(String[] args) {
    //加载配置文件
ApplicationContext applicationContext =
            new ClassPathXmlApplicationContext("bean.xml");
    //通过 bean 生成 User 对象
User user = (User)applicationContext.getBean("user");
    //输出结果
System.out.println(user.toString());
}
```

7.4.2 自动装配

set 注入和构造注入有时在做配置时比较麻烦。所以框架为了提高开发效率，提供自动装配功能，简化配置。Spring 框架是默认不支持自动装配的，要想使用自动装配需要修改 Spring 配置文件中<bean>标签的 autowire 属性。

Spring 的<bean>标签中包含一个 autowire 属性，我们可以通过设置 autowire 的属性值来自动装配 Bean。所谓自动装配，就是将一个 Bean 自动地注入其他 Bean 的 Property 中。

autowire 属性有 5 个值，其值及说明如表 7-2 所示。

表 7-2 autowire 属性值

属性值	说明
default(默认值)	由<bean>的上级标签<beans>的 default-autowire 属性值确定。例如<beans default-autowire="byName">，则该<bean>元素中的 autowire 属性对应的属性值就为 byName
byName	根据属性的名称自动装配。容器将根据名称查找与属性完全一致的 Bean，并将其属性自动装配
byType	根据属性的数据类型(Type)自动装配，如果一个 Bean 的数据类型兼容另一个 Bean 中属性的数据类型，则自动装配
constructor	根据构造函数参数的数据类型，进行 byType 模式的自动装配
no	在默认情况下，不使用自动装配，Bean 依赖必须通过 ref 元素定义

在 User 类中为 Cat 自定义属性赋值，可以使用自动装配来完成。

```
<bean id="cat" class="com.yyzy.xml.Cat">
<property name="cname" value="小花猫"></property>
<property name="age" value="3"></property>
</bean>
<bean id="user" class="com.yyzy.xml.User" autowire="byName">
<property name="username" value="李宁"></property>
<property name="sex" value="男"></property>
<property name="age" value="23"></property>
</bean>
```

7.4.3　Spring 注解方式

在 Spring 中，尽管使用 XML 配置文件可以实现 Bean 的装配工作，但如果应用中有很多 Bean 时，会导致 XML 配置文件过于臃肿，给后续的维护和升级工作带来一定的困难。为此，Spring 提供了对 Annotation（注解）技术的全面支持。

Spring 2.5 提供了一个 context 命名空间，它提供了通过扫描类包以及应用注解定义 Bean 的方式。它为我们引入了组件扫描机制，在类路径下寻找标注了"@Component、@Service、@Controller"等注解的类，并把这些类纳入 Spring 容器中管理。

要使用自动扫描机制，我们需要增加以下配置信息：

（1）Spring 配置文件增加 context 命名空间与 xsd 的引用，使用注解，需要 context 组件解析配置文件。

（2）增加 context 的动态扫描。

```
<beans xmlns="http://www.springframework.org/schema/beans"
xmlns:xsi="http://www.w3.org/2001/XMLSchema-instance"
xmlns:context="http://www.springframework.org/schema/context"
xsi:schemaLocation="http://www.springframework.org/schema/beans
        http://www.springframework.org/schema/beans/spring-beans-4.3.xsd
        http://www.springframework.org/schema/context
    http://www.springframework.org/schema/context/spring-context-4.3.xsd">
<context:component-scan base-package="com.yyzy.xml"></context:component-scan>
```

Spring 中定义了一系列的注解，常用的注解可以分为组件注解和属性注解两部分。

组件注解：

@Component：可以使用此注解描述 Spring 中的 Bean，但它是一个泛化的概念，仅仅表示一个组件（Bean），并且可以作用在任何层次。使用时只需将该注解标注在相应类上即可。

@Repository：用于将数据访问层（DAO 层）的类标识为 Spring 中的 Bean，其功能与 @Component 相同。

@Service：通常作用在业务层（Service 层），用于将业务层的类标识为 Spring 中的 Bean，其功能与 @Component 相同。

@Controller：通常作用在控制层（如 Spring MVC 的 Controller），用于将控制层的类标识为 Spring 中的 Bean，其功能与 @Component 相同。

属性注解：

@Value(value=""）：用于基本数据类型。

@Resource(name=""）：用于引用类型。

@Autowired：根据数据类型自动扫描并装配。

在实际项目中，基本类型通常通过从界面传递值或者从数据库读取，而对于引用类型，经常使用 @Autowired 注解。

对前面的 Cat 类和 User 类使用注解方式。

```
@Component
public class Cat{
    @Value("小白")
    String cname;
    @Value("3")
    int age;
}
@Controller("user")
public class User{
    @Value("张珊珊")
    private String username;
    @Value("女")
    private String sex;
    @Value("23")
    private    Integer age;
    @Autowired
    Cat cat;
}
```

本章小结

本章主要介绍了依赖注入和控制反转概念，介绍了 Bean 的配置以及 Bean 实例化的三种方式。

用户可以通过属性注入的方式建立 Bean 和 Bean 的依赖，也可以通过构造函数的方式完成任务。本章还介绍了自动装配以及注解的使用。

课后习题

1. 请简述什么是 Spring 的 IoC 和 DI。
2. 请简述 Bean 的几种装配方式的基本用法。

第8章 Spring 之 AOP

 学习目标

[本章知识点]
　　AOP 含义
　　AOP 术语
[思政目标]
　　使学生理解团队协作的重要性，了解团队合作是计算机应用程序开发小组成员所遵循的基本规范之一。
　　使学生理解分类学的思想是人类解决复杂问题时最常用的方法之一，在学习生活中做好分类计划，合理规划时间。

8.1 AOP

"AOP"为"Aspect Oriented Programming"的缩写，意为面向切面编程，是通过预编译方式和运行期动态代理，实现程序功能统一维护的一种技术。

AOP 是 OOP（面向对象编程）的延续，是软件开发中的一个热点，也是 Spring 框架中的一个重要内容，是函数式编程的一种衍生范型。

利用 AOP 可以对业务逻辑的各个部分进行隔离，从而使得业务逻辑各部分之间的耦合度降低，提高程序的可重用性，同时提高了开发的效率。AOP 采取横向抽取机制，取代了传统纵向继承体系重复性代码。

8.2　AOP 术语

（1）target（目标类）：需要被代理的类。例如：UserService。
（2）Joinpoint（连接点）：指那些可能被拦截到的方法。例如：所有的方法。
（3）PointCut（切入点）：已经被增强的连接点。例如：addUser()。
（4）advice（通知/增强处理）：增强代码。例如：after、before。
（5）Weaving（织入）：指把增强 advice 应用到目标对象（target）来创建新的代理对象（proxy）的过程。
（6）proxy（代理类）：将通知应用到目标对象之后，被动态创建的对象。
（7）Aspect（切面）：是切入点 pointcut 和通知 advice 的结合。
一根线是一个特殊的面；一个切入点和一个通知组成一个特殊的面。
各个术语解释如图 8-1 所示。

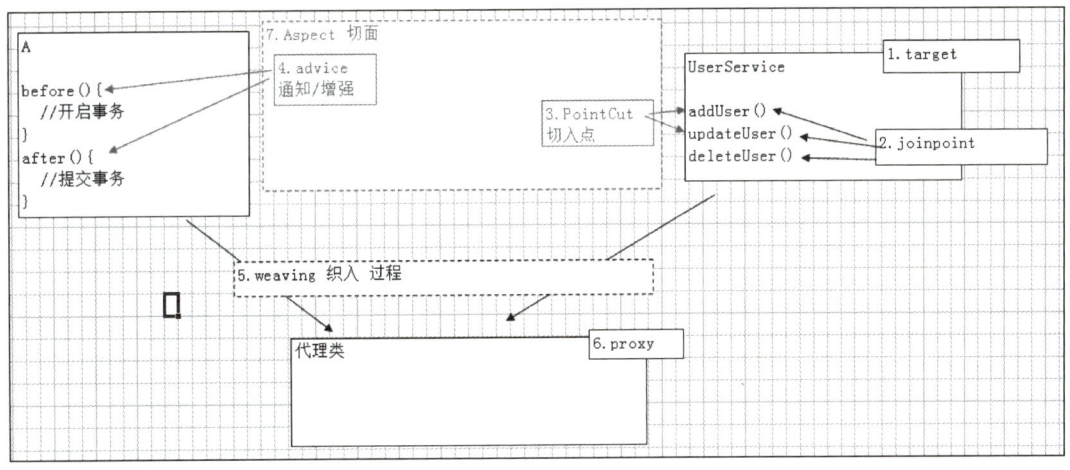

图 8-1　AOP 术语解释

8.3　AOP 配置及实现

8.3.1　Spring 的通知类

AOP 联盟为通知 Advice 定义了 org.aopalliance.aop.Advice。按照通知 Advice 在目标类方法的连接点位置，Advice 可以分为 5 类：

● 前置通知 org.springframework.aop.BeforeAdvice，在目标方法执行前实施增强。

- 后置通知 org.springframework.aop.AfterReturningAdvice，在目标方法执行后实施增强。
- 环绕通知 org.aopalliance.intercept.MethodInterceptor，在目标方法执行前后实施增强。
- 异常抛出通知 org.springframework.aop.ThrowsAdvice，在方法抛出异常后实施增强。
- 最终通知 org.springframework.aop.AfterAdvice，在目标组件的方法正常执行后执行，或在异常通知之前执行。

8.3.2 实现 AOP 的方式一：基于 XML 的声明式

基于 XML 的声明式是指通过 XML 文件来定义切面、切入点及通知，所有的切面、切入点和通知都必须定义在<aop:config>元素内。<aop:config>元素及其子元素如图 8-2 所示。

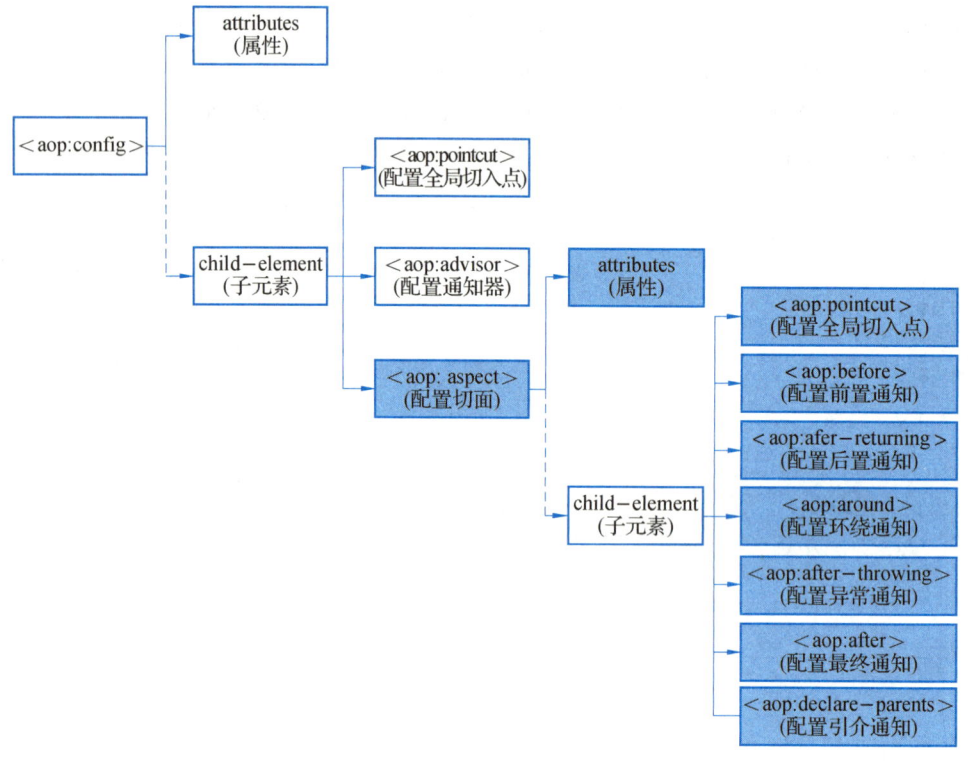

图 8-2 <aop:config>元素及其子元素

Spring 配置文件中的<beans>元素下可以包含多个<aop:config>元素，一个<aop:config>元素中又可以包含属性和子元素，其子元素包括<aop:pointcut>、<aop:advisor>和<aop:aspect>。在配置时，这 3 个子元素必须按照此顺序来定义。在子元素下，同样包含了属性和多个子元素，通过使用<aop:aspect>元素及其子元素就可以在 XML 文件中配置切面、切入点和通知。

1. 配置切面

在 Spring 的配置文件中，配置切面使用的是<aop:aspect>元素，该元素会将一个已定义好的 SpringBean 转换成切面 Bean，所以要在配置文件中先定义一个普通的 SpringBean（如上述代码中定义的 myAspect）。定义完成后，通过<aop:aspect>元素的 ref 属性即可引用该 Bean。

第 8 章 Spring 之 AOP

配置<aop:aspect>元素时,通常会指定 id 和 ref 两个属性,如表 8-1 所示。

表 8-1 <aop:aspect>元素的属性及其描述

属性名称	描述
id	用于定义该切面的唯一标识名称
ref	用于引用普通的 SpringBean

2. 配置切入点

在 Spring 的配置文件中,切入点是通过<aop:pointcut>元素来定义的。当<aop:pointcut>元素作为<aop:config>元素的子元素定义时,表示该切入点是全局切入点,它可被多个切面所共享;当<aop:pointcut>元素作为<aop:aspect>元素的子元素时,表示该切入点只对当前切面有效。

在定义<aop:pointcut>元素时,通常会指定 id 和 expression 两个属性,如表 8-2 所示。

表 8-2 <aop:pointcut>元素的属性及其描述

属性名称	描述
id	用于定义该切面的唯一标识名称
expression	用于指定切入点关联的切入点表达式

切入点表达式,用于声明 Spring 容器中哪些组件的函数是目标函数,也就是切面程序要作用到哪些组件的哪些函数上。

语法:execution(修饰符 返回值 包.类.方法名(参数) throws 异常)

案例 1 如下所示。

```
execution(*com.yyzy.crm.*.service..*.*(..))
```

案例 2 如下所示。

```
<aop:pointcut expression="execution(*com.yyzy.crm.service.*.*(..)) | |
execution(*com.yyzy.*Do.*(..))" id="myPointCut"/>
```

3. 配置通知

在配置代码中,分别使用子元素配置了 5 种常用通知,这 5 个子元素不支持使用子元素,但在使用时可以指定一些属性,如表 8-3 所示。

表 8-3 通知的常用属性及其描述

属性名称	描述
pointcut	该属性用于指定一个切入点表达式,Spring 将在匹配该表达式的连接点时植入该通知
pointcut-ref	该属性用于指定一个已经存在的切入点名称。通常 pointcut 和 pointcut-ref 两个属性只需要使用其中一个

续表

属性名称	描述
method	该属性用于指定一个方法名，指定将切面 Bean 中的该方法转换为增强处理
returning	该属性只对〈after-returning〉元素有效，它用于指定一个形参名，后置通知方法可以通过形参访问目标方法的返回值

下面通过一个案例来说明 AOP 的配置及应用。

(1) 创建 Maven 工程 chapter8，添加 pom.xml 依赖（spring-context，aspectjweaver，junit）。

```
<dependencies>
<dependency>
<groupId>org.springframework</groupId>
<artifactId>spring-context</artifactId>
<version>5.1.7.RELEASE</version>
</dependency>
<dependency>
<groupId>org.aspectj</groupId>
<artifactId>aspectjweaver</artifactId>
<version>1.9.7</version>
</dependency>
<dependency>
<groupId>junit</groupId>
<artifactId>junit</artifactId>
<version>4.12</version>
<scope>compile</scope>
</dependency>
</dependencies>
```

(2) 在 Chapter8 项目的 src 下的 Java 目录下，创建一个 com.yyzy.crm.service 包，在该包中创建目标类 UserDao。

```
public class UserDao {
    public void addUser() {
        System.out.println("添加用户");
    }
    public void deleteUser() {
        System.out.println("删除用户");
    }
}
```

(3) 在 Chapter8 项目的 src 下的 Java 目录下，创建一个 com.yyzy.crm.aspect 包，在该包中创建切面类 MyAspect，并在类中分别定义不同类型的通知。

```
public class MyAspect {
    //前置通知
    public void myBefore(JoinPoint joinPoint) {
```

```java
        System.out.print("----前置通知----");
    }
    //后置通知
    public void myAfterReturning(JoinPoint joinPoint){
        System.out.print("----后置通知----");
    }
    //环绕通知
    public Object myAround(ProceedingJoinPoint proceedingJoinPoint)
            throws Throwable{
        System.out.println("-----环绕开始------");
        //执行当前目标方法
        Object obj=proceedingJoinPoint.proceed();
        //结束
        System.out.println("-----环绕结束-----");
        return obj;
    }
    //异常通知
    publicvoidmyAfterThrowing(JoinPoint joinPoint,Throwable e){
        System.out.println("异常通知:"+"出错了"+e.getMessage());
    }
    //最终通知
    public void myAfter(){
        System.out.println("----最终通知----");
    }
}
```

(4)在 resources 下面创建 bean.xml 里面写切面配置。

```xml
<beans xmlns="http://www.springframework.org/schema/beans"
xmlns:xsi="http://www.w3.org/2001/XMLSchema-instance"
xmlns:aop="http://www.springframework.org/schema/aop"
xsi:schemaLocation="http://www.springframework.org/schema/beans
        http://www.springframework.org/schema/beans/spring-beans-4.3.xsd
        http://www.springframework.org/schema/aop
        http://www.springframework.org/schema/aop/spring-aop-4.3.xsd">
<!--目标类-->
<bean id="userDao" class="com.yyzy.crm.UserDao"></bean>
<!--切面类-->
<bean id="myAspect" class="com.yyzy.crm.MyAspect"></bean>
<aop:config>
<!--配置切面-->
<aop:aspect ref="myAspect">
<!--配置切入点-->
<aop:pointcut id="mypoint" expression="execution(* com.yyzy.crm.service.UserDao.*(..))"/>
<!--配置前置通知-->
<aop:before method="myBefore" pointcut-ref="mypoint"></aop:before>
<!--配置后置通知-->
<aop:after-returning method="myAfterReturning" pointcut-ref="mypoint"></aop:after-returning>
```

```xml
<!--配置环绕通知-->
<aop:around method="myAround" pointcut-ref="mypoint"></aop:around>
<!--配置异常通知-->
<aop:after-throwing method="myAfterThrowing" pointcut-ref="mypoint" throwing="e"></aop:after-throwing>
<!--配置最终通知-->
<aop:after method="myAfter" pointcut-ref="mypoint"></aop:after>
</aop:aspect>
</aop:config>
</beans>
```

(5) 在 com.yyzy.crm.aspect 包下创建测试类 TestAspect。

```java
public static void main(String args[]) {
ApplicationContext applicationContext = new ClassPathXmlApplicationContext("bean.xml");
    //从 spring 容器获得内容
UserDao userDao = (UserDao) applicationContext.getBean("userDao");
    //执行方法
userDao.addUser();
userDao.deleteUser();
    }
```

8.3.3 实现 AOP 的方式二：基于注解的声明式

常用的注解如表 8-4 所示。

表 8-4 注解及其描述

注解名称	描述
@Aspect	用于定义一个切面
@Pointcut	用于定义切入点表达式。在使用时还需定义一个包含名字和任意参数的方法签名来表示切入点名称。实际上，这个方法签名就是一个返回值为 void，且方法体为空的普通的方法
@Before	用于定义前置通知，相当于 BeforeAdvice。在使用时，通常需要指定一个 value 属性值，该属性值用于指定一个切入点表达式（可以是已有的切入点，也可以直接定义切入点表达式）
@AfterReturning	用于定义后置通知，相当于 AfterReturningAdvice。在使用时可以指定 pointcut-value 和 returning 属性，其中 pointcut-value 这两个属性的作用一样，都用于指定切入点表达式。returning 属性值用于表示 Advice 方法中可定义与此同名的形参，该形参可用于访问目标方法的返回值
@Around	用于定义环绕通知，相当于 MethodInterceptor。在使用时需要指定一个 value 属性，该属性用于指定该通知被植入的切入点

续表

注解名称	描述
@AfterThrowing	用于定义异常通知来处理程序中未处理的异常，相当于 ThrowAdvice。在使用时可指定 pointcut-value 和 throwing 属性。其中 pointcut-value 用于指定切入点表达式，而 throwing 属性值用于指定一个形参名来表示 Advice 方法中可定义与此同名的形参，该形参可用于访问目标方法抛出的异常
@After	用于定义最终 final 通知，不管是否异常，该通知都会执行。使用时需要指定一个 value 属性，该属性用于指定该通知被植入的切入点

下面通过一个例子来说明注解的使用。

（1）在 Chapter8 项目的 src 的 Java 目录下，创建一个 com.yyzy.crm.annotation 包，在该包中创建切面类 MyAspect2。

```
@Component
@Aspect
public class MyAspect2{
    //定义切入点表达式
    @Pointcut("execution(* com.yyzy.crm.UserDao.*(..))")
    //使用一个返回值为 void、方法体为空的方法来命名切入点
    private void myPointCut(){}
    //前置通知
    @Before("myPointCut()")
    public void myBefore(JoinPoint joinPoint){
        System.out.print("----前置通知-----");
    }
    //后置通知
    @AfterReturning("myPointCut()")
    public void myAfterReturning(JoinPoint joinPoint){
        System.out.print("----后置通知----");
    }
    //环绕通知
    @Around("myPointCut()")
    public Object myAround(ProceedingJoinPoint proceedingJoinPoint)
            throws Throwable{
        System.out.println("-----环绕开始------");
        //执行当前目标方法
        Object obj = proceedingJoinPoint.proceed();
        //结束
        System.out.println("-----环绕结束-----");
        return obj;
    }
    //异常通知
    @AfterThrowing(value = "myPointCut()",throwing = "e")
    public void myAfterThrowing(JoinPoint joinPoint, Throwable e){
        System.out.println("异常通知:"+"出错了" + e.getMessage());
```

```
        }
        //最终通知
        @After("myPointCut()")
        public void myAfter(){
System.out.println("----最终通知----");
        }
    }
```

(2)在目标类中添加注解。

```
@@Service("user")
public class UserDao{
    public void addUser(){
System.out.println("添加用户");
    }
    public void deleteUser(){
System.out.println("删除用户");
    }
}
```

(3)创建配置文件 bean2.xml,并对该文件进行编辑。

```
<beans xmlns="http://www.springframework.org/schema/beans"
xmlns:xsi="http://www.w3.org/2001/XMLSchema-instance"
xmlns:aop="http://www.springframework.org/schema/aop"
xmlns:context="http://www.springframework.org/schema/context"
xsi:schemaLocation="http://www.springframework.org/schema/beans
    http://www.springframework.org/schema/beans/spring-beans-4.3.xsd
    http://www.springframework.org/schema/aop
    http://www.springframework.org/schema/aop/spring-aop-4.3.xsd
    http://www.springframework.org/schema/context
    http://www.springframework.org/schema/context/spring-context-4.3.xsd">
<!--指定需要扫描的包,使注解生效 -->
<context:component-scan base-package="com.yyzy.crm" />
<!--启动基于注解的声明式 AspectJ 支持 -->
<aop:aspectj-autoproxy />
</beans>
```

(4)测试。

```
public static void main(String args[]){
ApplicationContext applicationContext = new ClassPathXmlApplicationContext("bean2.xml");
        //从 spring 容器获得内容
UserDao userDao = (UserDao) applicationContext.getBean("user");
        //执行方法
userDao.addUser();
userDao.deleteUser();
    }
```

第8章 Spring 之 AOP

本章小结

　　本章主要讲解 AOP 的概念和作用，要求学生理解 AOP 中的相关常用术语，熟悉 Spring 中 AOP 的配置和应用，并能够掌握 AOP 的基于 XML 和注解的两种开发方式。

课后习题

1. 请列举 AOP 专业术语并解释。
2. 请列举 Spring 的通知类型并解释。

第 9 章 MyBatis 与 Spring 的整合

学习目标

[本章知识点]
　　整合 jar 包介绍
　　IoC 容器配置
　　基于 MapperScannerConfigurer 的整合
　　利用 Springtest 模块测试

[思政目标]
　　使学生理解沟通能力是一个人生存与发展的必备技能，了解软件开发过程中沟通能力的重要性，学习高效沟通的三大秘诀。

前面章节分别讲解了 Spring 和 MyBatis 的相关知识，然而在实际的项目开发中，Spring 与 MyBatis 都是整合在一起使用的。在读者掌握了 MyBatis 的使用后，本章将对 MyBatis 与 Spring 的整合内容进行详细讲解。

9.1 整合 jar 包介绍

Spring 模块包含 10 个 jar 包，如图 9-1 所示。

图 9-1　Spring 模块 jar 包

第 9 章 MyBatis 与 Spring 的整合

MyBatis 核心 jar 包 1 个加上 Spring 整合 MyBatis jar 包 1 个，如图 9-2 所示。

```
org.mybatis:mybatis:3.2.8
org.mybatis:mybatis-spring:1.2.2
```

图 9-2 MyBatis 核心 jar 包

MySQL 连接 jar 包 1 个，如图 9-3 所示。

```
mysql:mysql-connector-java:5.1.35
```

图 9-3 MySQL 连接 jar 包

日志包 3 个，如图 9-4 所示。

```
log4j:log4j:1.2.17
org.slf4j:slf4j-api:1.7.12
org.slf4j:slf4j-log4j12:1.7.12
```

图 9-4 日志包

单元测试包 1 个，如图 9-5 所示。

```
junit:junit:4.13-beta-2
```

图 9-5 单元测试包

pom.xml 文件依赖如下所示。

```xml
<?xmlversion="1.0" encoding="UTF-8"?>
<project xmlns="http://maven.apache.org/POM/4.0.0"
xmlns:xsi="http://www.w3.org/2001/XMLSchema-instance"
xsi:schemaLocation="http://maven.apache.org/POM/4.0.0 http://maven.apache.org/xsd/maven-4.0.0.xsd">
<modelVersion>4.0.0</modelVersion>
<groupId>org.test</groupId>
<artifactId>SpringMyBatisDemo</artifactId>
<version>1.0-SNAPSHOT</version>
<properties>
<maven.compiler.source>8</maven.compiler.source>
<maven.compiler.target>8</maven.compiler.target>
<project.build.sourceEncoding>UTF-8</project.build.sourceEncoding>
<spring.version>4.0.2.RELEASE</spring.version>
<mybatis.version>3.2.8</mybatis.version>
<slf4j.version>1.7.12</slf4j.version>
<log4j.version>1.2.17</log4j.version>
</properties>
<dependencies>
<!-- Spring 框架包 start -->
<dependency>
```

```xml
    <groupId>org.springframework</groupId>
    <artifactId>spring-test</artifactId>
    <version>${spring.version}</version>
</dependency>
<dependency>
    <groupId>org.springframework</groupId>
    <artifactId>spring-core</artifactId>
    <version>${spring.version}</version>
</dependency>
<dependency>
    <groupId>org.springframework</groupId>
    <artifactId>spring-oxm</artifactId>
    <version>${spring.version}</version>
</dependency>
<dependency>
    <groupId>org.springframework</groupId>
    <artifactId>spring-tx</artifactId>
    <version>${spring.version}</version>
</dependency>
<dependency>
    <groupId>org.springframework</groupId>
    <artifactId>spring-jdbc</artifactId>
    <version>${spring.version}</version>
</dependency>
<dependency>
    <groupId>org.springframework</groupId>
    <artifactId>spring-aop</artifactId>
    <version>${spring.version}</version>
</dependency>
<dependency>
    <groupId>org.springframework</groupId>
    <artifactId>spring-context</artifactId>
    <version>${spring.version}</version>
</dependency>
<dependency>
    <groupId>org.springframework</groupId>
    <artifactId>spring-context-support</artifactId>
    <version>${spring.version}</version>
</dependency>
<dependency>
    <groupId>org.springframework</groupId>
    <artifactId>spring-expression</artifactId>
    <version>${spring.version}</version>
</dependency>
<dependency>
    <groupId>org.springframework</groupId>
    <artifactId>spring-orm</artifactId>
    <version>${spring.version}</version>
```

```xml
        </dependency>
        <!-- Spring 框架包 end -->
        <!--MyBatis 框架包 start -->
        <dependency>
            <groupId>org.mybatis</groupId>
            <artifactId>mybatis</artifactId>
            <version>${mybatis.version}</version>
        </dependency>
        <dependency>
            <groupId>org.mybatis</groupId>
            <artifactId>mybatis-spring</artifactId>
            <version>1.2.2</version>
        </dependency>
        <!--MyBatis 框架包 end -->
        <!--数据库驱动 -->
        <dependency>
            <groupId>mysql</groupId>
            <artifactId>mysql-connector-java</artifactId>
            <version>5.1.35</version>
        </dependency>
        <!--数据库驱动-->
        <!-- log start -->
        <dependency>
            <groupId>log4j</groupId>
            <artifactId>log4j</artifactId>
            <version>${log4j.version}</version>
        </dependency>
        <dependency>
            <groupId>org.slf4j</groupId>
            <artifactId>slf4j-api</artifactId>
            <version>${slf4j.version}</version>
        </dependency>
        <dependency>
            <groupId>org.slf4j</groupId>
            <artifactId>slf4j-log4j12</artifactId>
            <version>${slf4j.version}</version>
        </dependency>
        <!-- log END -->
        <!--test jar-->
        <dependency>
            <groupId>junit</groupId>
            <artifactId>junit</artifactId>
            <version>4.13-beta-2</version>
        </dependency>
        <!--test jar-->
    </dependencies>
</project>
```

本次学习案例数据库代码如下。

```
USE 'mybatisdb';
DROP TABLE IF EXISTS 'users';
CREATE TABLE 'users' (
  'id' int(11) NOT NULL auto_increment,
  'name' varchar(10) NOT NULL,
  'pwd' varchar(11) NOT NULL,
  'seachCol' text,
  PRIMARY KEY ('id')
) ENGINE=InnoDB DEFAULT CHARSET=utf8;
```

项目结构如图 9-6 所示。

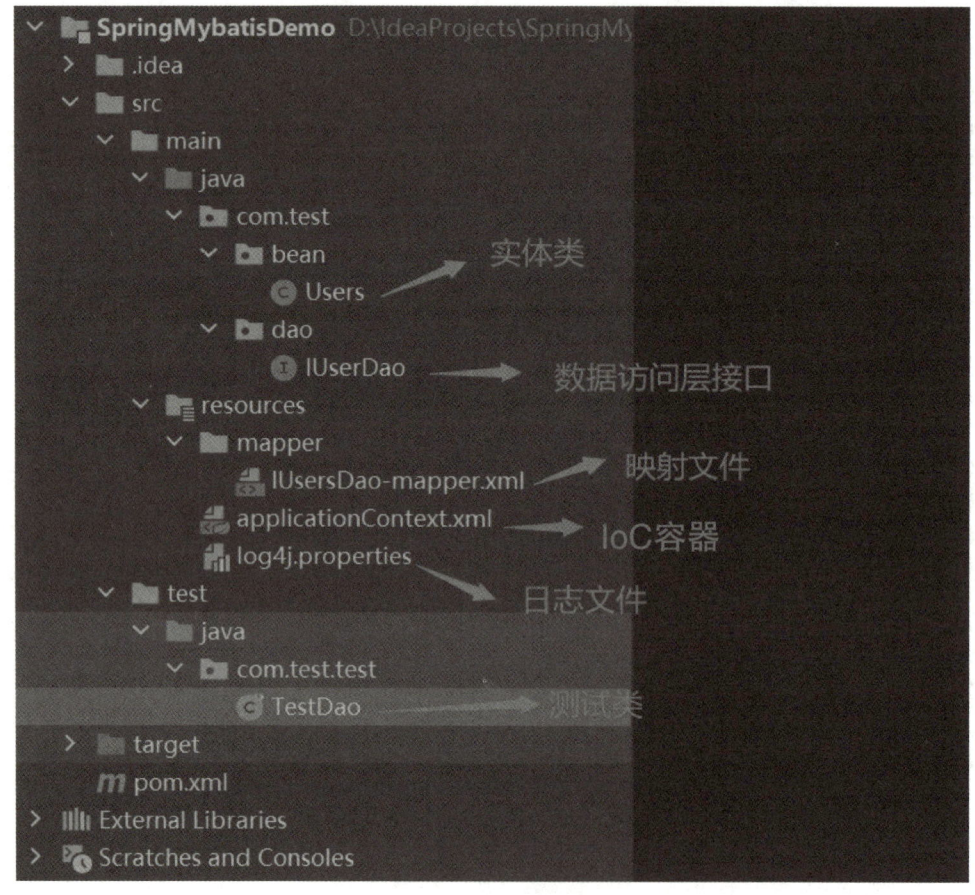

图 9-6　项目结构图

9.2 案 例

9.2.1 编写实体类

在 com.test.bean 下根据 users 表创建实体类(Users.java)。

```
package com.test.bean;

public class Users{
private Integer id;
private String name;
private String pwd;

public Users(){
super();
//TODO Auto-generated constructor stub
}
public Users(Integer id,String name,String pwd){
super();
this.id = id;
this.name = name;
this.pwd = pwd;
}
public Integer getId(){
return id;
}
public void setId(Integer id){
this.id = id;
}
public String getName(){
return name;
}
public void setName(String name){
this.name = name;
}
public String getPwd(){
return pwd;
}
public void setPwd(String pwd){
this.pwd = pwd;
}

}
```

9.2.2 编写数据访问层接口

在 com.test.dao 下创建数据访问层接口(IUserDao.java)，并且定义基本 CURD 操作方法。

```java
package com.test.dao;

import java.util.List;

import com.test.bean.Users;

public interface IUserDao{
/**添加用户信息*/
public Integer insertUser(Users users) throws Exception;
/** 根据主键删除用户信息 */
public Integer delUserByUid(Integer uid) throws Exception;
/**根据主键修改用户信息 */
public Integer updateUserByUid(Users users) throws Exception;
/**根据主键查询用户信息*/
public Users getUserByUid(Integer uid) throws Exception;
/**查询所有用户信息*/
public List<Users> getUserList() throws Exception;
/**查询用户总数据量*/
public long getUserCount() throws Exception;
}
```

9.2.3 编写映射文件

在 resources 下创建 mapper 文件夹，并且在 mapper 文件夹中创建数据访问层接口的映射文件，为接口中的方法配置 SQL 语句。

```xml
<!DOCTYPE mapper PUBLIC
"-//mybatis.org//DTD Mapper 3.0//EN"
    "http://mybatis.org/dtd/mybatis-3-mapper.dtd">
<mapper namespace="com.test.dao.IUserDao">
<insert id="insertUser" parameterType="com.test.bean.Users">
    insert into users(name,pwd) values (#{name},#{pwd})
</insert>
<delete id="delUserByUid" parameterType="int">
    delete from users where id=#{id}
</delete>
<update id="updateUserByUid" parameterType="com.test.bean.Users">
    update users set name=#{name},pwd=#{pwd} where id=#{id}
</update>
<select id="getUserByUid" parameterType="int" resultType="com.test.bean.Users">
    select * from users where id=#{id}
```

```xml
</select>
<select id="getUserList" resultType="com.test.bean.Users">
    select * from users
</select>
<select id="getUserCount" resultType="long">
    select count(*) from users
</select>
</mapper>
```

9.2.4 IoC 容器配置

在 resources 下创建 IoC 容器(applicationContext.xml),用于管理工程中的 Bean 对象。

```xml
<beans xmlns="http://www.springframework.org/schema/beans"
xmlns:xsi="http://www.w3.org/2001/XMLSchema-instance"
xmlns:context="http://www.springframework.org/schema/context"
xsi:schemaLocation="http://www.springframework.org/schema/beans
    http://www.springframework.org/schema/beans/spring-beans-4.0.xsd
    http://www.springframework.org/schema/context
    http://www.springframework.org/schema/context/spring-context-4.0.xsd">
<context:component-scan base-package="com.test.service"></context:component-scan>
<!--1.配置数据源 DriverManagerDataSource -->
<bean id="ds" class="org.springframework.jdbc.datasource.DriverManagerDataSource">
    <property name="driverClassName" value="com.mysql.jdbc.Driver"></property>
    <property name="url" value="jdbc:mysql://localhost:3306/mybatisdb"></property>
    <property name="username" value="root"></property>
    <property name="password" value="root"></property>
</bean>
<!--2.配置 SqlSessionFactoryBean
dataSource:数据源
mapperLocations:MyBatis 的映射文件
-->
<bean id="ssf" class="org.mybatis.spring.SqlSessionFactoryBean">
    <property name="dataSource" ref="ds"></property>
    <property name="mapperLocations" value="classpath:mapper/*-mapper.xml"></property>
</bean>
<!--3.为数据访问层接口 配置一个扫描器 MapperScannerConfigurer
去扫描接口并且以动态代理的形式为接口生成实例对象
sqlSessionFactory:依赖于 SqlSessionFactoryBean 的实例
basePackage:为哪个包结构下的接口生成动态代理对象-->
<bean id="msc" class="org.mybatis.spring.mapper.MapperScannerConfigurer">
    <!--<property name="sqlSessionFactory" ref="ssf"></property>-->
    <property name="sqlSessionFactoryBeanName" value="ssf"></property>
    <property name="basePackage" value="com.test.dao"></property>
</bean>
</beans>
```

（1）DriverManagerDataSource 类为 Spring 中提供的一个数据源驱动管理类，主要用于连接数据库。

driverClassName 属性：用于配置数据库驱动。

url 属性：用于配置数据库连接字符串。

username 属性：用于配置数据库连接用户名。

password 属性：用于配置数据库连接密码。

（2）SqlSessionFactoryBean 类为 MyBatis 中提供的一个工厂类，用于配置 MyBatis 以及创建 SqlSession。

dataSource 属性：用于配置数据源。

mapperLocations 属性：用于配置映射文件。

以下为 SqlSessionFactoryBean 类中属性。

```java
public classSql SessionFactoryBean implements FactoryBean<SqlSessionFactory>, InitializingBean, ApplicationListener<ApplicationEvent> {
    private static final Log logger = LogFactory.getLog(SqlSessionFactoryBean.class);
    private Resource configLocation;
    private Resource[] mapperLocations;
    private DataSource dataSource;
    private TransactionFactory transactionFactory;
    private Properties configurationProperties;
    private SqlSessionFactoryBuilder sqlSessionFactoryBuilder = new SqlSessionFactoryBuilder();
    private SqlSessionFactory sqlSessionFactory;
    private String environment = SqlSessionFactoryBean.class.getSimpleName(); // EnvironmentAware requires spring 3.1
    private booleanfailFast;
    private Interceptor[] plugins;
    private TypeHandler<?>[] typeHandlers;
    private String typeHandlersPackage;
    private Class<?>[] typeAliases;
    private String typeAliasesPackage;
    private Class<?>typeAliasesSuperType;
    private DatabaseIdProvider databaseIdProvider; // issue #19. No default provider.
    private ObjectFactory objectFactory;
    private ObjectWrapperFactory objectWrapperFactory;
}
```

configLocation 属性：配置 MyBatis 原生配置文件。

plugins 属性：配置 MyBatis 插件。

typeAliases/typeAliasesPackages 属性：用于配置别名。

MapperScannerConfigurer 类：为 MyBatis 提供的一个映射扫描配置类，MapperScannerConfigurer 是为了解决 MapperFactoryBean 的繁琐而产生的，有了 MapperScannerConfigurer 就不需要我们为每个映射接口声明一个 Bean 了，大大缩减了开发的效率。

basePackage 属性：配置映射接口的包，包里面的所有的接口将被扫描到。

sqlSessionFactoryBeanName 属性：用于注入 sqlSessionFactory。

9.3 利用 Springtest 模块进行测试

在 test 下创建测试类（TestDao.java）用于测试数据访问层。

```java
package com.test.test;
import com.test.bean.Users;
import com.test.dao.IUserDao;
import org.junit.Test;
import org.junit.runner.RunWith;
import org.springframework.beans.factory.annotation.Autowired;
import org.springframework.test.context.ContextConfiguration;
import org.springframework.test.context.junit4.SpringJUnit4ClassRunner;
@RunWith(SpringJUnit4ClassRunner.class)
@ContextConfiguration("classpath:applicationContext.xml")
public class TestDao{
@Autowired
private IUserDao dao;
@Test
public void testInsertUser(){
try{
Users u=new Users(null,"corn","123456");
int i=dao.insertUser(u);
System.out.println(">>>"+i);
}catch(Exception e){
e.printStackTrace();
}
}
@Test
public void testDelUserByUid(){
try{
int i=dao.delUserByUid(103);
System.out.println(">>>"+i);
}catch(Exception e){
e.printStackTrace();
}
}
@Test
public void testUpdateUserByUid(){
try{
Users u=new Users(102,"corn","123456");
int i=dao.updateUserByUid(u);
System.out.println(">>>"+i);
}catch(Exception e){
e.printStackTrace();
```

```
            }
        }
    @Test
    public void testGetUserByUid(){
        try{
            Users u=dao.getUserByUid(102);
            System.out.println(u);
        }catch(Exception e){
            e.printStackTrace();
        }
    }
    @Test
    public void testGetUserList(){
        try{
            dao.getUserList().forEach(u->System.out.println(u));
        }catch(Exception e){
            e.printStackTrace();
        }
    }
    @Test
    public void testGetUserCount(){
        try{
            Long count=dao.getUserCount();
            System.out.println(count);
        }catch(Exception e){
            e.printStackTrace();
        }
    }
}
```

（1）@RunWith(SpringJUnit4ClassRunner.class)：用于指定 Junit 运行环境，是 Junit 提供给其他框架测试环境接口扩展，为了便于使用 Spring 的依赖注入，Spring 提供了 SpringJUnit4ClassRunner 作为 Junit 测试环境。

（2）@ContextConfiguration("classpath:applicationContext.xml")：导入配置文件。

（3）@ContextConfiguration({"classes=Congfig.clsss","classpath：applicationContext.xml"})：使用 classes 来直接导入同包下写的配置类，或者导入配置文件。

本章小结

本章首先对 MyBatis 与 Spring 框架整合的环境搭建进行了讲解，然后讲解了基于 Mapper 接口方式的开发整合流程。通过本章的学习，读者应能够熟练地掌握 MyBatis 与 Spring 框架的整合方式，这将为后面项目的学习打下坚实的基础。

课后习题

1. 请简述 MyBatis 与 Spring 整合所需 jar 包的种类。
2. 请简述 MapperFactoryBean 和 MapperScannerConfigurer 的作用。

第10章 Spring MVC 框架

学习目标

[本章知识点]
　　Spring MVC 概述
　　Spring MVC 框架后台工作原理
　　Spring MVC 的工作流程
[思政目标]
　　让学生养成多读书、读好书的习惯，做一个有知识、有才学、有作为的年轻人。

10.1 Spring MVC 简介

Spring MVC 是 Spring 框架提供的一个实现 Web MVC 设计模式的轻量级框架，它与 Sturts2 框架一样，都属于 MVC 框架。

（1）三层架构。

表现层：Web 层，用来和客户端进行数据交互，表现层一般会采用 MVC 的设计模式。

业务层：处理公司具体的业务逻辑。

持久层：用来操作数据库。

（2）MVC 的设计模型。

MVC 的全称是"Model View Controller"，即模型视图控制器，每个部分都各司其职。

Model：数据模型，实体类对象，用来进行数据封装。

View：视图，具体指的是 JSP、HTML 等语言编写的视图，用来展示数据给用户看。

Controller：整个流程的控制器，用来接收用户的请求，用来进行数据校验。

Spring MVC 是 Spring 提供的一个强大而灵活的 Web 框架。借助于注解，Spring MVC 提供

了几乎和 POJO 一样的开发模式,使得控制器的开发和测试更加简单。这些控制器一般不直接处理请求,而是将其委托给 Spring 上下文中的其他 Bean,通过 Spring 的依赖注入功能,这些 Bean 被注入控制器中。

Spring MVC 主要由 DispatcherServlet、处理器映射(找控制器)、适配器(调用控制器的方法)、控制器(业务)、视图解析器、视图组成。

10.2 Spring MVC 的优点

Spring MVC 是 Spring 提供的一个实现了 Web MVC 设计模式的轻量级 Web 框架。它与 Struts2 框架一样,都属于 MVC 框架,但其在使用和性能等方面比 Struts2 更加优异。

Spring MVC 具有如下特点。

- 是 Spring 框架的一部分,可以方便地利用 Spring 所提供的其他功能。
- 灵活性强,易于与其他框架集成。
- 提供了一个前端控制器 DispatcherServlet,使开发人员无须额外开发控制器对象。
- 可自动绑定用户输入,并能正确地转换数据类型。
- 内置了常见的校验器,可以校验用户输入。如果校验不能通过,那么就会重定向到输入表单。
- 支持国际化。可以根据用户区域显示多国语言。
- 支持多种视图技术。它支持 JSP、Velocity、FreeMarker 等视图技术。
- 使用基于 XML 的配置文件,在编辑后不需要重新编译应用程序。

10.3 Spring MVC 框架工作原理

Spring MVC 在项目中具体是怎么执行的呢? Spring MVC 程序的执行情况如图 10-1 所示。按照图 10-1 中所标注的序号,Spring MVC 程序的完整执行流程如下。

(1)用户通过客户端向服务器发送请求,请求会被 Spring MVC 的前端控制器 DispatcherServlet 所拦截。

(2)DispatcherServlet 拦截到请求后,会调用 HandlerMapping 映射处理器。

(3)映射处理器根据请求 URL 找到具体的处理器,生成处理器对象及处理器拦截器(如果有则生成)一并返回给 DispatcherServlet。

(4)DispatcherServlet 会根据返回信息选择合适的 HandlerAdapter(处理器适配器)。

(5)HandlerAdapter 会调用并执行 Handler(处理器),这里的处理器指的就是程序中编写

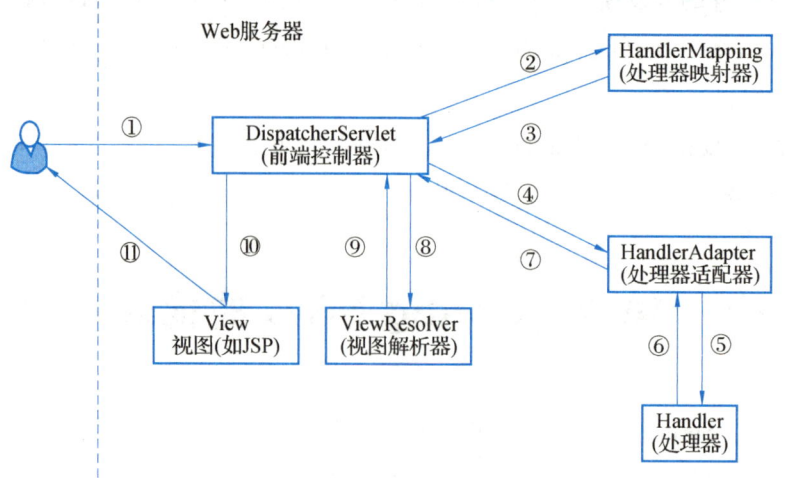

图 10-1　Spring MVC 的工作原理

的 Controller 类，也被称之为后端控制器。

（6）Handler 执行完成后，会返回一个 ModelAndView 对象，该对象中会包含视图名或包含模型和视图名。

（7）HandlerAdapter 将 ModelAndView 对象返回给 DispatcherServlet。

（8）DispatcherServlet 会根据 ModelAndView 对象选择一个合适的 ViewResolver（视图解析器）。

（9）ViewResolver 解析后，会向 DispatcherServlet 中返回具体的 View（视图）。

（10）DispatcherServlet 对 View 进行渲染（将模型数据填充至视图中）。

（11）视图渲染结果会返回客户端浏览器显示。

在上述执行过程中，DispatcherServlet、HandlerMapping、HandlerAdapter 和 ViewResolver 对象的工作是在框架内部执行的，开发人员并不需要关心这些对象内部的实现过程，只需要配置前端控制器（DispatcherServlet），完成 Handler 中的业务处理，并在视图（View）中展示相应信息即可。

涉及的处理器如下。

DispatcherServlet：前端处理器，用户请求到达前端控制器。它就相当于 MVC 模式中的 C。DispatcherServlet 是整个流程控制的中心，由它调用其他组件处理用户的请求，DispatcherServlet 的存在降低了组件之间的耦合性。

HandlerMapping：映射处理器。HandlerMapping 负责根据用户请求找到 Handler 即处理器，Spring MVC 提供了不同的映射器实现不同的映射方式，例如配置文件方式、实现接口方式、注解方式等。

HandlerAdapter：适配器处理器。通过 HandlerAdapter 对处理器进行执行，这是适配器模式的应用，通过扩展适配器可以对更多类型的处理器进行执行。

ViewResolver：视图解析器。ViewResolver 负责将处理结果生成 View 视图，ViewResolver

首先根据逻辑视图名解析成物理视图名即具体的页面地址,再生成 View 视图对象,最后对 View 进行渲染将处理结果通过页面展示给用户。Spring MVC 框架提供了很多的 View 视图类型,包括 jstlView、freemarkerView、pdfView 等。

10.4 Spring MVC 核心类

10.4.1 DispatcherServlet

DispatcherServlet 的全名是"org.springframework.web.servlet.DispatcherServlet",它在程序中充当着前端控制器的角色。在 Java Web 中后台使用 Servlet 进行调度时,使用 HttpServletRequest 获取请求,使用 HttpServletResponse 进行响应。在 Spring MVC 框架中原理同样如此,只不过不再像 Web 那样,每个 Controller 都要继承 HttpServlet 那么麻烦,而是整个后台使用一个 DispatcherServlet,该 Servlet 在 web.xml 配置文件中进行配置,具体含义如表 10-1 所示。

表 10-1 web.xml 中配置 DispatcherServlet 属性

属性	描述
servlet-name	默认查找 Spring MVC 配置文件 xxx-servlet.xml
servlet-class	DispatcherServlet 类的全路径(org.springframework.web.servlet.DispatcherServlet)
url-pattern	匹配拦截器要拦截的文件类型
init-param	可以通过把属性 param-name 的值设置成 contextConfigLocation 来配置 Spring MVC 配置文件的位置。如果不配置该参数,默认到 web/WEB-INF/下查找 springMVC-servlet.xml

具体代码如下所示。

```
<web-app version="2.4"
xmlns="http://java.sun.com/xml/ns/j2ee"
xmlns:xsi="http://www.w3.org/2001/XMLSchema-instance"
xsi:schemaLocation="http://java.sun.com/xml/ns/j2ee http://java.sun.com/xml/ns/j2ee/web-app_2_4.xsd">
<servlet>
<display-name>DispatcherServlet</display-name>
<servlet-name>springMVC</servlet-name>
<servlet-class>org.springframework.web.servlet.DispatcherServlet</servlet-class>
<!--
<init-param>
```

```
<param-name>contextConfigLocation</param-name>
<param-value>classpath:springMVC-servlet.xml</param-value>
</init-param>
    -->
</servlet>
<servlet-mapping>
<servlet-name>springMVC</servlet-name>
<url-pattern>/</url-pattern>
</servlet-mapping>
</web-app>
```

10.4.2 HandlerMapping

映射处理器 HandlerMapping 是请求映射处理器，也就是通过请求的 URL 找到对应的逻辑处理单元，有很多实现类：SimpleUrlHandlerMapping、BeanNameUrlHandlerMapping、ControllerClassNameHandlerMapping。其中，SimpleUrlHandlerMapping 思路清晰；BeanNameUrlHandlerMapping 是系统默认，匹配跟 URL 同名的数据对象，可以省略。所以，省略配置 HandlerMapping，在配置 Controller 时使用跟 URL 同名的 id 即可。

事实上，当我们不配置 HandlerMapping 的时候，Spring MVC 框架默认会帮我们配置 BeanNameUrlHandlerMapping，该处理器的作用是生成跟 URL 同名的数据对象。比如，我们请求地址是 login.action，该处理器就会生成一个同名的数据对象 login.action。所以当我们不配置 HandlerMapping 的时候，是 Spring MVC 框架已经帮助我们搭建好了 BeanNameUrlHandlerMapping，可以直接使用。

10.4.3 ViewResolver

Spring MVC 用于处理视图最重要的两个接口是 ViewResolver 和 View。ViewResolver 的主要作用是把一个逻辑上的视图名称解析为一个真正的视图，Spring MVC 中把 View 对象呈现给客户端的是 View 对象本身，而 ViewResolver 只是把逻辑视图名称解析为对象的 View 对象。View 接口的主要作用是用于处理视图，然后返回给客户端。

常见的 ViewResolver 有 AbstractCachingViewResolver、UrlBasedViewResolver、InternalResourceViewResolver、XmlViewResolver、BeanNameViewResolver、ResouceBundleViewResolver。

在 Spring MVC 中可以同时定义多个 ViewResolver 视图解析器，然后它们会组成一个 ViewResolver 链。当 Controller 处理器方法返回一个逻辑视图名称后，ViewResolver 链将根据其中 ViewResolver 的优先级来进行处理。所有的 ViewResolver 都实现了 Ordered 接口，在 Spring 中实现了这个接口的类都可以排序。在 ViewResolver 中是通过 order 属性来指定顺序的，默认是最大值，值越小优先级越大。InternalResourceViewResolver 能解析所有的视图，即永远能返回一个非空 View 对象的 ViewResolver，因此一定要把它放在 ViewResolver 链的最后面。

10.4.4　Controller

Controller 是用户自定义类，是后台服务器的核心部分，通过 HandlerAdapter 查找到。

AbstractController 是简单的控制器，继承该类的方法需要实现 handleRequestInternal 方法，该方法有两个参数（request 和 response），返回值是 ModelAndView 类型的对象。该方法原理与普通的 Servlet 相同。

MultiActionController 控制器可以将多个请求处理方法合并在一个控制器里，这样可以把相关功能组合在一起。

使用注解@RequestMapping，可以直接根据 url 匹配方法，简单明了。

10.4.5　ModelAndView

使用 ModelAndView 类来存储处理完后的结果数据，以及显示该数据的视图。从名字上看"ModelAndView"中的"Model"代表模型，"View"代表视图，这个名字就很好地解释了该类的作用。业务处理器调用模型层处理完用户请求后，把结果数据存储在该类的 model 属性中，把要返回的视图信息存储在该类的 view 属性中，然后让该 ModelAndView 返回该 Spring MVC 框架。框架通过调用配置文件中定义的视图解析器，对该对象进行解析，最后把结果数据显示在指定的页面上。

ModelAndView 的具体作用：

（1）返回指定页面。ModelAndView 构造方法可以指定返回的页面名称，也可以通过 setViewName( ) 方法跳转到指定的页面。

（2）返回所需数值。使用 ModelAndView 的 addObject( ) 方法设置需要返回的值，addObject( ) 有几个不同参数的方法，可以默认或指定返回对象的名字。

10.5　第一个 Spring MVC 程序

（1）使用 Maven 新建项目，选择模板 maven-archetype-webapp，搭建 Spring MVC 项目，如图 10-2 所示。

（2）在 pom.xml 中添加依赖 spring-context、spring-web、spring-webmvc。Spring-context 是 Spring 框架的依赖包，Spring MVC 框架需要引入两个依赖包：spring-web 和 spring-mvc。还会应用到 HttpServletRequest 等类，需要引入 Servlet 的 jar 包，所以还需要添加 servlet-api 的依赖。

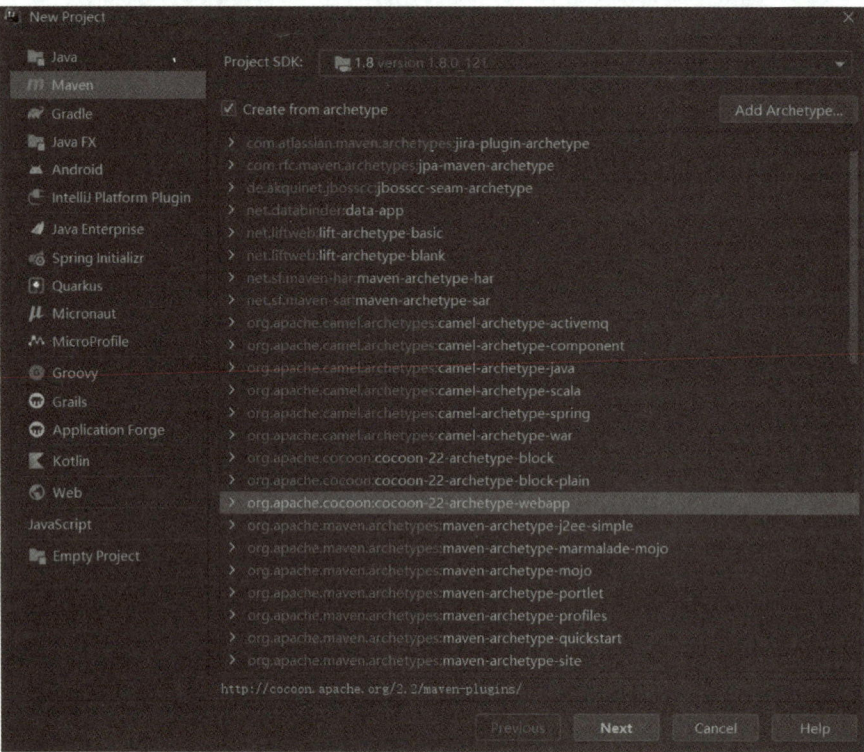

图 10-2　选择 Maven 模板

```
<dependencies>
<dependency>
<groupId>org.springframework</groupId>
<artifactId>spring-context</artifactId>
<version>5.1.7.RELEASE</version>
</dependency>
<dependency>
<groupId>org.springframework</groupId>
<artifactId>spring-web</artifactId>
<version>5.1.7.RELEASE</version>
</dependency>
<dependency>
<groupId>org.springframework</groupId>
<artifactId>spring-webmvc</artifactId>
<version>5.1.7.RELEASE</version>
</dependency>
<dependency>
<groupId>javax.servlet</groupId>
<artifactId>servlet-api</artifactId>
<version>2.5</version>
</dependency>
</dependencies>
```

第 10 章　Spring MVC 框架

（3）在 web.xml 文件中配置 DispatcherServlet。

```xml
<web-app version="2.4"
xmlns="http://java.sun.com/xml/ns/j2ee"
xmlns:xsi="http://www.w3.org/2001/XMLSchema-instance"
xsi:schemaLocation="http://java.sun.com/xml/ns/j2eehttp://java.sun.com/xml/ns/j2ee/web-app_2_4.xsd">
<servlet>
<display-name>DispatcherServlet</display-name>
<servlet-name>springMVC</servlet-name>
<servlet-class>org.springframework.web.servlet.DispatcherServlet</servlet-class>
<init-param>
<param-name>contextConfigLocation</param-name>
<param-value>classpath:springmvc-config.xml</param-value>
</init-param>
</servlet>
<servlet-mapping>
<servlet-name>springMVC</servlet-name>
<url-pattern>/</url-pattern>
</servlet-mapping>
</web-app>
```

主要对<servlet>元素进行配置。在<servlet>中，配置 Spring MVC 的前端控制器 DispatcherServlet，并通过其子元素<init-param>配置 Spring MVC 配置文件的位置；在<servlet-mapping>中，通过<url-pattern>元素的"/"，将所有 URL 拦截，并交由 DispatcherServlet 处理。

在<init-param>中参数 contextConfigLocation 指定 resources 下的配置文件，如果有多个使用","隔开；参数 classpath 表示 resources root 的位置。

如果<init-param>元素存在并且通过其子元素配置了 Spring MVC 配置文件的路径，则应用程序在启动时会加载配置路径下的配置文件；如果没有通过<init-param>元素配置，则应用程序会默认到 WEB-INF 目录下寻找如下方式命名的配置文件。

servletName-servlet.xml

其中，"servletName"指的是部署在 web.xml 中的 DispatcherServlet 的名称，在上面 web.xml 中的配置代码中即为 springmvc，而"-servlet.xml"是配置文件名的固定写法，所以应用程序会在 WEB-INF 下寻找 springmvc-servlet.xml。

（4）创建 Controller 类。在 Java 文件夹下创建 com.yyzy.controller 包，在该包中创建控制器类 FirstController，该类需要实现 Controller 接口。

```java
public class FirstController implements Controller{
    public ModelAndView handleRequest(HttpServletRequest httpServletRequest, HttpServletResponse httpServletResponse) throws Exception{
        //创建 ModelAndView 对象
        ModelAndView mav = new ModelAndView();
```

```
        //向模型对象中添加数据
mav.addObject("这是我的第一个Spring MVC程序");
        //设置逻辑视图名
mav.setViewName("/WEB-INF/jsp/first.jsp");
        //返回ModelAndView对象
        return mav;
    }
}
```

handleRequest()是Controller接口的实现方法,FirstController类会调用该方法来处理请求,并返回一个包含视图名或包含视图名和模型的ModelAndView对象。本案例中,向模型对象中添加了一个名称为msg的字符串对象,并设置返回的视图路径为"/WEB-INF/jsp/first.jsp",这样请求就会被转发到first.jsp页面。

(5)Spring框架配置文件。

Spring MVC框架本是Spring框架的一部分,所以可以和Spring框架共用一个配置文件,但是命名方式应与Spring MVC框架的命名方式一致。

在resources下面创建Spring MVC框架配置文件springmvc-config.xml。

```
<!--配置处理器Handle,映射到firstController-->
<bean name="/firstController" class="com.yyzy.controller.FirstController"></bean>
<!--配置处理器映射器,将处理器handle的name作为URL进行查找-->
<bean class="org.springframework.web.servlet.mvc.SimpleControllerHandlerAdapter"></bean>
<!--配置视图解析器-->
<bean class="org.springframework.web.servlet.view.InternalResourceViewResolver"></bean>
```

首先定义一个名称为"FirstController"的Bean,该Bean会将控制器类FirstController映射到"/firstController"请求中;然后配置映射处理器BeanNameUrlHandlerMapping和处理器适配器SimpleControllerHandlerAdapter,其中映射处理器用于将处理器Bean中的name(即url)进行处理器查找,而处理器适配器用于完成对FirstController处理器中handleRequest()方法的调用。最后配置视图解析器InternalResourceViewResolver来解析结果视图,并将结果呈现给用户。

(6)创建视图页面。

在WEB-INF目录下,创建一个jsp文件夹,并在文件夹中创建一个页面文件first.jsp,在该页面中使用EL表达式获取msg中的信息。

```
<%@ page contentType="text/html;charset=UTF-8" language="java" %>
<html>
<head>
<title>Title</title>
</head>
<body>
欢迎大家来到我的课堂!<br>
${msg}
</body>
</html>
```

第 10 章　Spring MVC 框架

（7）启动项目，测试应用。

运行 Web 项目，需要在 IDEA 中配置 Tomcat。找到 apache-tomcat 文件夹，把本地 Tomcat 路径链接到 IDEA 的 Tomcat 插件下，让其可以工作，然后在 IDEA 下面新建 Tomcat，最后把当前项目部署到 Tomcat 服务器上。

在 IDEA 的 run 目录上找到 Edit Configurations，弹出如图 10-3 所示对话框。

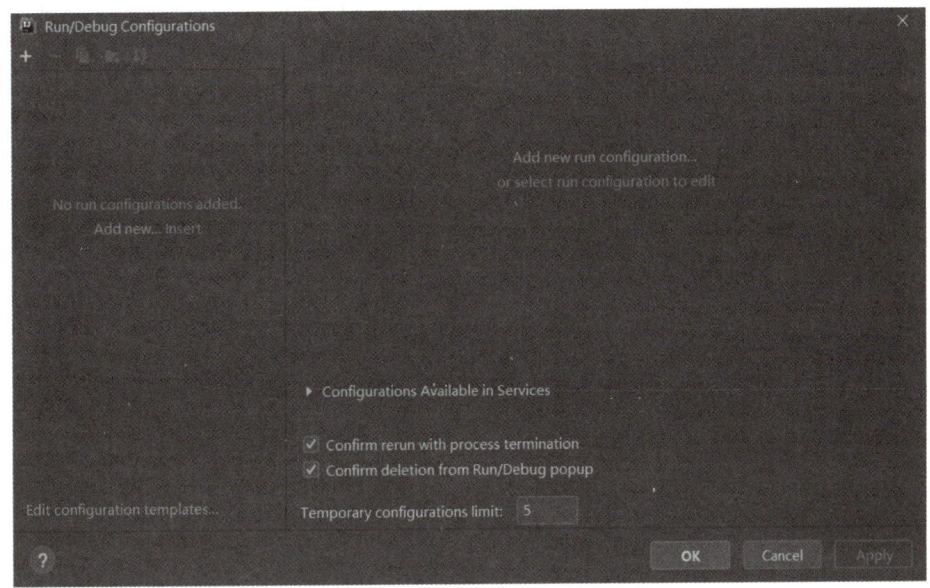

图 10-3　配置 Tomcat 对话框

点击"+"，找到 Tomcat Server，选择 Local，如图 10-4 所示。

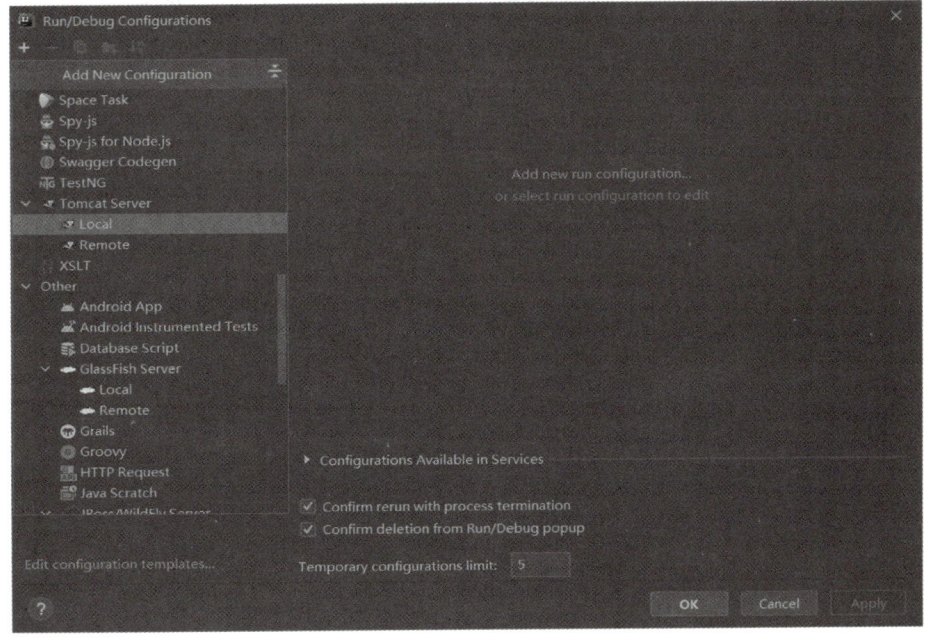

图 10-4　链接到 Tomcat 本地

点 Configuration，弹出对话框，选择本地 Tomcat 文件夹，如图 10-5 所示。

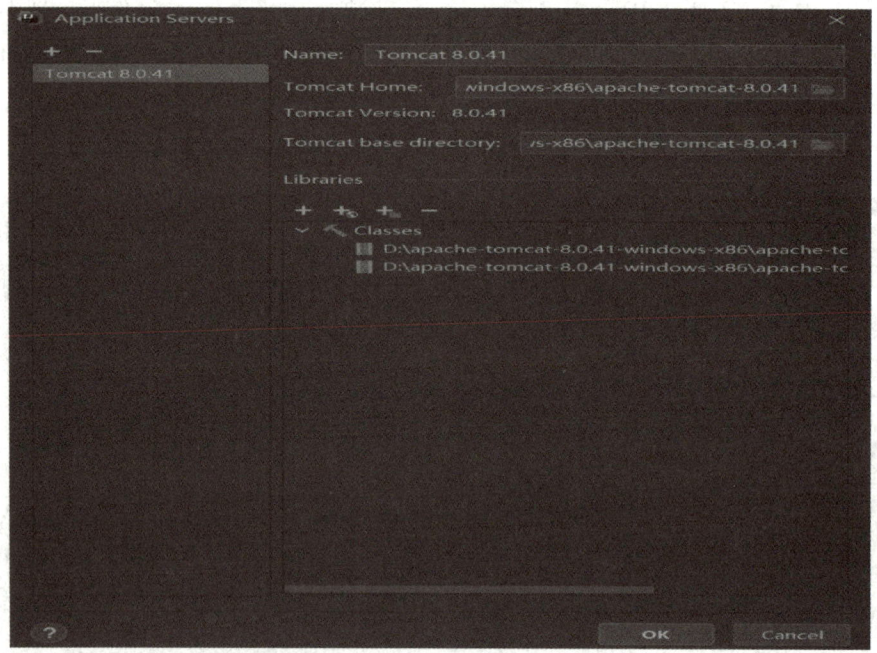

图 10-5　配置 Tomcat 文件夹

点击"OK"后，对话框完成，点击"Apply"，Tomcat 插件配置完成，弹出如图 10-6 所示画面。下面要在 IDEA 中新建 Tomcat 并部署本项目。

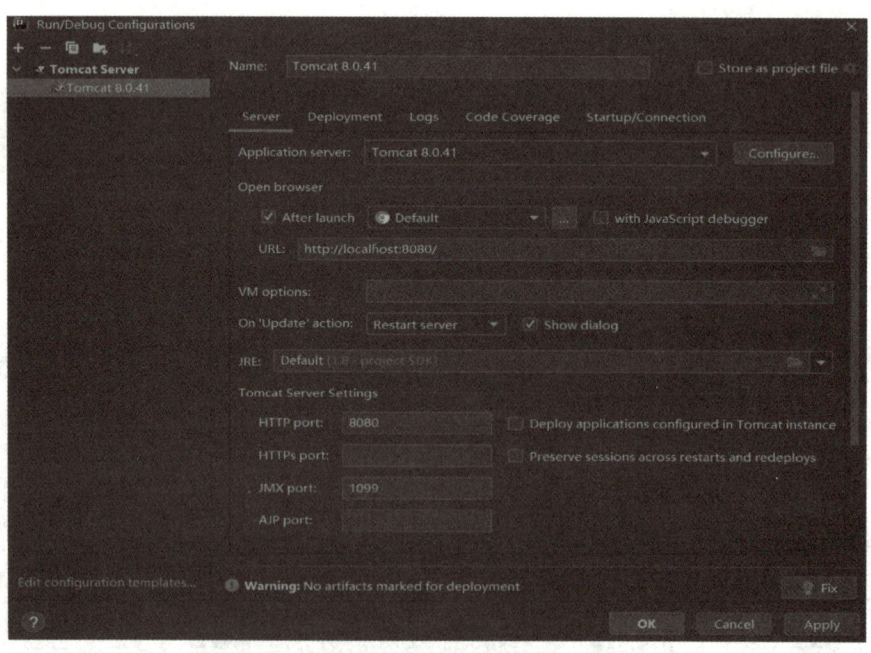

图 10-6　Tomcat 上部署本项目

点击 Deployment 按钮，添加项目到服务器上，点击"+"添加项目，如图 10-7 所示。

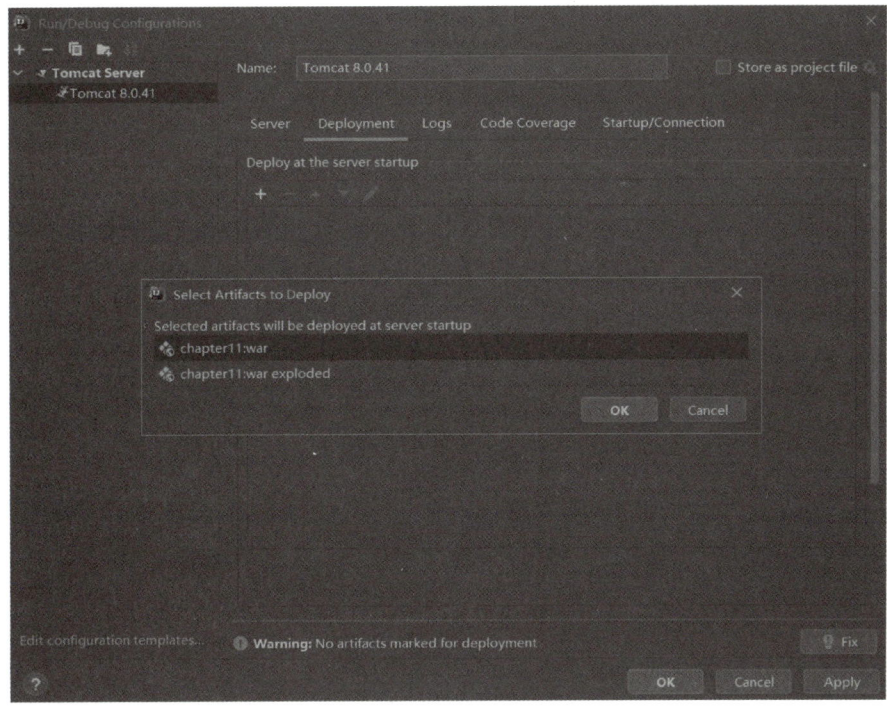

图 10-7　添加项目

点击"OK"后，在 Tomcat 上就可以看到添加的项目，如图 10-8 所示。

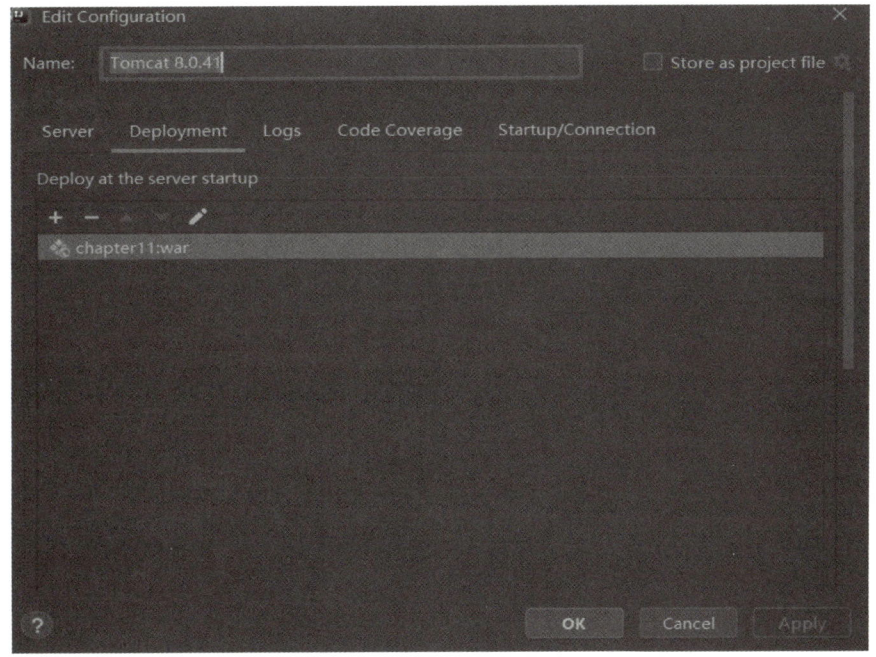

图 10-8　项目添加完成

启动服务器，运行项目，在浏览器中输入 http://localhost:8080/chapter11/firstController 这个地址，即可看到运行结果显示到页面中了。

本章小结

　　本章首先对 Spring MVC 框架进行了简单的介绍，然后讲解了 DispatcherServlet 核心类，最后通过入门案例对 Spring MVC 的工作流程进行了详细讲解。通过本章的学习，读者能够了解什么是 Spring MVC 以及 Spring MVC 的优点，掌握 Spring MVC 入门程序的编写，并熟悉 Spring MVC 框架的工作流程。

课后习题

1. 请简述 Spring MVC 框架的工作流程。
2. 请简述 Controller 注解的使用步骤。
3. 请列举请求处理方法的参数类型和返回类型(至少 5 个)。

第 11 章
Spring MVC 框架注解和数据传递

 学习目标

[本章知识点]
　　Spring MVC 框架常用注解
　　Spring MVC 框架的数据传递
[思政目标]
　　引导学生关心国家大事，了解国家最新政策，增强爱党、爱国的情感。
　　引导学生重视中国文化，继承和弘扬中华优秀传统文化。

11.1 Spring MVC 框架常用注解

Spring MVC 可以使用注解，需要使用注解时在配置文件中配置注解扫描即可。

```
<!--配置注解扫描位置 -->
<context:component-scan base-package="com.yyzy.controller"/>
```

base-package：需要指向控制类所在的包。

常用注解：

@Controller：用于指示 Spring 类的实例是一个控制器。

@RequestMapping：用于把 URL 与具体方法进行匹配。

@RequstParam：绑定单个请求数据，可以是 URL 中的数据、表单提交的数据或上传的文件。

11.2　Controller 注解类型

org. springframework. stereotype. Controller 注解类型用于指示 Spring 类的实例是一个控制器，其注解形式为@ Controller。该注解在使用时不需要再实现 Controller 接口，只需要将@ Controller 注解加入到控制器类上，然后通过 Spring 的扫描机制找到标注了该注解的控制器即可。

```
@ Controller
public class UserController {

}
```

为了保证 Spring 能够找到控制器类，还需要在 Spring MVC 的配置文件中添加如上所示的扫描配置信息。

使用注解方式时，程序的运行需要依赖 Spring 的 AOP 包，因此需要添加 AOP 的依赖引入 AOP 的 jar 包，否则程序运行时会报错。

11.3　RequestMapping 注解类型

Spring 通过@ Controller 注解找到相应的控制器类后，还需要知道控制器内部对每一个请求是如何处理的，这就需要使用 RequestMapping 注解类型。RequestMapping 注解类型用于映射一个请求或一个方法，其注解形式为@ RequestMapping，可以将该注解标注在一个方法或一个类上。

（1）当标注在一个方法上时，该方法将成为一个请求处理方法，它会在程序接收到对应的 URL 请求时被调用。将@ RequestMapping 注解标注在方法上的示例如下。

```
@ Controller
public class UserController {
    @ RequestMapping(value = "first")
    public ModelAndView handleRequest(HttpServletRequest httpServletRequest, HttpServletResponse httpServletResponse) throws Exception {
        // 创建 ModelAndView 对象
        ModelAndView mav = new ModelAndView();
        return mav;
    }
}
```

使用@ RequestMapping 注解后，上述代码中的 handleRequest（ ）方法就可以通过地址 http://localhost:8080/chapter12/first 进行访问。

（2）标注在类上时，则表明访问此类路径下的方法都要加上其配置的路径。使用@RequestMapping 注解标注在类上的示例如下。

```
@Controller
@RequestMapping(value = "/user")
public class UserController {
    @RequestMapping(value = "first")
    public ModelAndView handleRequest(HttpServletRequest httpServletRequest, HttpServletResponse httpServletResponse) throws Exception {
        // 创建 ModelAndView 对象
        ModelAndView mav = new ModelAndView();
        return mav;
    }
}
```

由于在类上添加了@RequestMapping 注解，并且其 value 属性值为"/user"，所以上述代码方法的请求路径将变为 http://localhost:8080/chapter12/user/first。如果该类中还包含其他方法，那么在其他方法的请求路径中也需要加入"/user"。

11.4 @RequestParam 注解

当前台表单元素的 name 属性和后台数据库的列名不一致时，可以使用@RequestParam 对应。如前端页面表单元素是 username，而后台 JavaBean 的属性是 uname，则接收不到参数，此时使用注解@RequestParam，可以使用不同名字的注解进行匹配。

```
PublicStringuserLogin(@RequestParam("username")String uname) {
}
```

11.5 请求处理方法的参数类型和返回类型

在控制器类中，每一个请求处理方法都可以有多个不同类型的参数，以及一个多种类型的返回结果。

在前面案例中，请求处理方法返回的是一个 ModelAndView 类型的数据。除了此种类型外，请求处理方法还可以返回其他类型的数据。Spring MVC 所支持的常见方法返回类型如下：

ModelAndView

Model

Map

View

String

void

其中 ModelAndView 类型中可以添加 Model 数据，并指定视图；String 类型的返回值可以跳转视图，但不能携带数据；而 void 类型主要在异步请求时使用，它只返回数据，而不会跳转视图。

由于 ModelAndView 类型未能实现数据与视图之间的解析，所以在企业开发时，方法的返回类型通常都会使用 String。既然 String 类型的返回值不能携带数据，那么在方法中是如何将数据带入视图页面的呢？这就用到了上面所讲解的 Model 参数类型，通过增加一个 Model 类型的参数，即可添加需要在视图中显示的属性。

```
@RequestMapping(value = "adduser")
    public String addUser(HttpServletRequest httpServletRequest, HttpServletResponse httpServletResponse, Model model){
    model.addAttribute("msg","hello,你还好吗?");
        return "WEB-INF/jsp/first";
    }
```

1. redirect 重定向

在修改用户信息操作后，将请求重定向到用户查询方法的实现代码如下所示。

```
@RequestMapping(value = "update")
    public String updateUser(HttpServletRequest httpServletRequest, HttpServletResponse httpServletResponse, Model model){
    //重定向到查询
        return "redirect:findUser";
    }
```

2. forward 请求转发

用户执行修改操作时，转发到用户修改页面的实现代码如下所示。

```
@RequestMapping(value = "edit")
    public String updateUser(HttpServletRequest httpServletRequest, HttpServletResponse httpServletResponse, Model model){
    //请求转发
        return "forward:editUser";
    }
```

11.6 视图解析器 ViewReSolver

Spring MVC 中的视图解析器负责解析视图。可以通过在配置文件中定义一个 ViewResolver 来配置视图解析器，在视图解析器中设置视图的前缀和后缀属性。这样设置后，方法中所定义的 view 路径将可以简化。例如，前面案例中的逻辑视图名只需设置为"first"，而不再需要设置为"/WEB-INF/jsp/first.jsp"，在访问时视图解析器会自动地增加前缀。

```xml
<beans xmlns="http://www.springframework.org/schema/beans"
xmlns:xsi="http://www.w3.org/2001/XMLSchema-instance"
xsi:schemaLocation="http://www.springframework.org/schema/beans
http://www.springframework.org/schema/beans/spring-beans-4.3.xsd">
<!--配置视图解析器-->
<bean class="org.springframework.web.servlet.view.InternalResourceViewResolver">
<property name="prefix" value="/WEB-INF/jsp/"></property>
<property name="suffix" value=".jsp"></property>
</bean>
</beans>
```

11.7 数据绑定

在执行程序时，Spring MVC 会根据客户端请求参数的不同，将请求消息中的信息以一定的方式转换并绑定到控制器类的方法参数中。这种将请求消息数据与后台方法参数建立连接的过程就是 Spring MVC 中的数据绑定。

1. 绑定简单数据类型

简单数据类型的绑定，就是指 Java 中几种基本数据类型的绑定，如 int、String、Double 等类型。

以用户登录页面为例。

```html
<form action="userLogin" method="post">
用户名:<input type="text" name="username"><br>
密码:<input type="text" name="password"><br>
<input type="submit" value="登录">
</form>
```

在控制器类 UserController 中增加 login() 方法，方法参数修改为使用简单数据类型的形式。

```
@RequestMapping("/userLogin")
    public String login(String username, String password){
System.out.println(username+","+password);
        return "success";
    }
```

前端页面请求中参数名和后台控制器类方法中的形参名不一样,这就会导致后台无法正确绑定并接收到前端请求的参数,需使用@RequestParam注解来进行间接数据绑定。

```
@RequestMapping("/userLogin")
    public String login(@RequestParam(value="username") String name, @RequestParam(value="password") String pwd){
System.out.println(username+","+password);
        return "success";
    }
```

2. 绑定对象

在使用简单数据类型绑定时,传递的参数个数有限。然而在实际应用中,客户端请求可能会传递多个不同类型的参数数据,那么就需要手动编写多个不同类型的参数,这种操作显然比较繁琐。此时就可以使用对象进行数据绑定。

对象数据绑定就是将所有关联的请求参数封装在一个对象中,然后在方法中直接使用该对象作为形参来完成数据绑定。

下面通过一个用户注册案例,来演示对象数据的绑定,具体实现步骤如下。

(1)在 Java 下创建包 com.yyzy.po,在包中创建 User 类。

```
public class User{
    int id;
    String username;
    String sex;
    int age;
get/set 和构造方法略
}
```

(2)在 webapp 下创建用户注册页面 register.jsp。

```
<form action="userRegister" method="post">
用户编号:<input type="text" name="id"><br>
用户姓名:<input type="text" name="username"><br>
用户性别:<input type="radio" name="sex" value="男">男
<input type="radio" name="sex" value="女">女<br>
用户年龄:<input type="text" name="age"><br>
<input type="submit" value="注册">
</form>
```

(3)在控制器类 UserController 中增加 userRegister()方法,方法参数修改为使用对象数据类型的形式。

第 11 章　Spring MVC 框架注解和数据传递

```
@RequestMapping("/userRegister")
public String userRegister(User user,Model model){
model.addAttribute("user",user);
    return "success";
}
```

userRegister()方法接收表单提交过来的信息时，应以 User 对象来接收。接收后通过 model 对象把数据保存到 user 中，跳转到 success.jsp 页面来显示。

```
<body>
用户姓名:${user.username}<br>
用户性别:${user.sex}<br>

用户年龄:${user.age}<br>
</body>
```

前端请求中，若有中文信息传递，如在用户名和性别输入框中，分别输入用户名"张三"和性别"男"时，虽然浏览器可以正确跳转到结果页面，但是在控制台中输出的中文信息却会出现乱码。这时可以使用 Spring 提供的编码过滤器来统一编码。要使用编码过滤器，只需要在 web.xml 中添加如下代码。

```
<!--配置编码过滤器-->
<filter>
<filter-name>CharacterEncodingFilter</filter-name>
<filter-class>
org.springframework.web.filter.CharacterEncodingFilter
</filter-class>
<init-param>
<param-name>encoding</param-name>
<param-value>UTF-8</param-value>
</init-param>
</filter>
<filter-mapping>
<filter-name>CharacterEncodingFilter</filter-name>
<url-pattern>/*</url-pattern>
</filter-mapping>
```

本章小结

本章主要讲解 Spring MVC 中的注解和数据绑定。首先讲解了三种基本注解，再讲解了简单数据类型、对象类型绑定。

课后习题

1. 请简述简单数据类型中的@RequestParam注解及其属性作用。
2. 请简述包装类型绑定时的注意事项。

第 12 章
JSON 数据交互和 RESTful 支持

[本章知识点]
　　JSON 概述
　　JSON 数据转换
　　RESTful 支持
　　应用案例——用户信息查询

[思政目标]
　　软件行业规划解析，培养学生的软件工匠精神。
　　引导学生认识到作为软件技术专业的一员，应更加明晰专业人才的培养目标，更加明确专业领域内工作岗位和工作内容的社会价值，自觉树立远大职业理想。

　　Spring MVC 在数据绑定的过程中，需要对传递数据的格式和类型进行转换，它既可以转换 String 类型的数据，也能够转换 JSON 等其他类型的数据。通过前面章节学习，读者已经掌握 String 等数据类型的转换和绑定，本章将针对 Spring MVC 中 JSON 类型的数据交互和 RESTful 支持进行详细讲解。

12.1　JSON 概述

　　JSON（JavaScript Object Notation，JS 对象标记）是一种轻量级的数据交换格式。它是基于 JavaScript 的一个子集，使用了 C、C++、C#、Java、JavaScript、Perl、Python 等其他语言的约束，采用完全独立于编程语言的文本格式来存储和表示数据。这些特性使 JSON 成为理想的数据交互语言，它易于阅读和编写，同时也易于机器解析和生成。
　　与 XML 一样，JSON 也是纯文本的数据格式。初学者可以使用 JSON 传输一个简单的

String、Number、Boolean，也可以传输一个数组或者复杂的 Object 对象。

12.1.1 对象结构

对象结构以"{"开始，以"}"结束，中间部分由 0 个或多个以英文逗号","分隔的 key/value 对构成（注意 key 和 value 之间以英文冒号":"分隔）。

```
{
    key1:value1,
    key2:value2,
    ...
}
```

其中 key 必须为 String 类型，value 可以是 String、Number、Object、Array 等数据类型。

12.1.2 数组结构

数组结构以"["开始，以"]"结束，中间部分由 0 个或多个以英文逗号","分隔的值的列表组成。

```
[
    value1,
    value2,
    ...
]
```

上述两种数据结构也可以分别组合构成更为复杂的数据结构。

12.2 JSON 数据转换

为了实现浏览器与控制器类（Controller）之间的数据交互，Spring 提供了一个 HttpMessageConverter<T>接口来完成此项工作，该接口主要用于将请求信息中的数据转化为一个类型为 T 的对象，并将类型为 T 的对象绑定到请求方法的参数中，或者将对象转换为响应信息传递给浏览器显示。

Spring 为 HttpMessageConverter<T>接口提供了很多实现类，这些实现类可以对不同类型的数据进行信息转换。其中 MappingJackson2HttpMessageConverter 是 Spring MVC 默认处理 JSON 格式请求响应的实现类。该实现类利用 Jackson 开源包读写 JSON 数据，可以将 Java 对象转换为 JSON 对象和 XML 文档，同时也可以将 JSON 对象和 XML 对象转换为 Java 对象。

要使用 MappingJackson2HttpMessageConverter 对数据进行转换，就需要使用 Jackson 的开源包，开发时所需的开源包如图 12-1 所示。

第 12 章　JSON 数据交互和 RESTful 支持

jackson-core-asl-1.9.7.jar	2013/2/4 11:20	Executable Jar File	224 KB
jackson-core-lgpl-1.9.7.jar	2013/2/4 11:20	Executable Jar File	229 KB
jackson-mapper-asl-1.9.7.jar	2013/2/4 11:20	Executable Jar File	763 KB
jackson-mapper-lgpl-1.9.7.jar	2013/2/4 11:20	Executable Jar File	768 KB

图 12-1　开源包

在使用注解式开发时，需要用到两个重要的 JSON 格式转换注解，分别为@ RequestBody 和@ ResponseBody，关于这两个注解的说明如表 12-1 所示。

表 12-1　@RequestBody 和@ResponseBody 注解

注解	说明
@ RequestBoby	用于将请求体中的数据绑定到方法的形参中。该注解用在方法的形参上
@ ResponseBody	用于直接返回 return 对象。该注解用在方法上

1. 项目讲解

（1）创建项目并导入相关 jar 包。使用 IDEA 创建一个名为 demo 的 Web 项目，然后将 Spring MVC 相关 jar 包、JSON 转换包依赖到 pom.xml 文件中。

（2）在 web.xml 中，对 Spring MVC 的前端控制器等信息进行配置。

```xml
<?xml version="1.0" encoding="UTF-8"?>
<web-app xmlns:xsi="http://www.w3.org/2001/XMLSchema-instance"
xmlns=http://java.sun.com/xml/ns/javaeexsi:schemaLocation="http://java.sun.com/xml/ns/javaee
http://java.sun.com/xml/ns/javaee/web-app_3_0.xsd"
id="WebApp_ID" version="3.0">
<display-name>demo</display-name>
<servlet>
    <servlet-name>springmvc</servlet-name>
    <servlet-class>
      org.springframework.web.servlet.DispatcherServlet
    </servlet-class>
    <init-param>
      <param-name>contextConfigLocation</param-name>
      <param-value>classpath:springmvc-config.xml</param-value>
    </init-param>
    <load-on-startup>1</load-on-startup>
</servlet>
<servlet-mapping>
    <servlet-name>springmvc</servlet-name>
    <url-pattern>/</url-pattern>
</servlet-mapping>
</web-app>
```

（3）在 src 目录下，创建 Spring MVC 的核心配置文件 springmvc-config.xml。

```xml
<?xml version="1.0" encoding="UTF-8"?>
<beans xmlns="http://www.springframework.org/schema/beans"
    xmlns:xsi="http://www.w3.org/2001/XMLSchema-instance"
    xmlns:mvc="http://www.springframework.org/schema/mvc"
    xmlns:context="http://www.springframework.org/schema/context"
    xmlns:tx="http://www.springframework.org/schema/tx"
    xsi:schemaLocation="http://www.springframework.org/schema/beans
    http://www.springframework.org/schema/beans/spring-beans.xsd
            http://www.springframework.org/schema/context
            http://www.springframework.org/schema/context/spring-context.xsd
            http://www.springframework.org/schema/mvc
            http://www.springframework.org/schema/mvc/spring-mvc.xsd
            http://www.springframework.org/schema/tx
            http://www.springframework.org/schema/tx/spring-mvc.xsd">
    <context:component-scan base-package="com.test.controller"/>
    <mvc:annotation-driven/>
    <mvc:resources location="/js/" mapping="/js/**"/>
    <bean id="viewResolver"
        class="org.springframework.web.servlet.view.InternalResourceViewResolver">
        <property name="prefix" value="/WEB-INF/jsp/"/>
        <property name="suffix" value=".jsp"/>
    </bean>
</beans>
```

不仅配置组件扫描器和视图解析器，还配置 Spring MVC 的注解驱动和静态资源访问映射。其中 `<mvc:annotation-driven/>` 配置会自动注册 RequestMappingHandlerMapping 和 RequestMappingHandlerAdapter 两个 Bean，并提供对读写 XML 和读写 JSON 等功能的支持。`<mvc:resources location="/js/" mapping="/js/**"/>` 元素用于配置静态资源的访问路径。由于在 web.xml 中配置的"/"会将页面中引入的静态文件也进行拦截，而拦截后页面中将找不到这些静态资源文件，这样就会引起页面报错。而增加了静态资源的访问映射配置后，程序就会自动地去配置路径下找静态内容。

`<mvc:resources location="/js/" mapping="/js/**"/>` 中有两个重要属性 location 和 mapping，关于这两个属性的说明如表 12-2 所示。

表 12-2　location 和 mapping 属性

属性	说明
location	用于定位需要访问的本地静态资源文件路径，具体到某个文件夹
mapping	匹配静态资源全路径，其中"/**"表示文件夹及其子文件夹下的某个具体文件

（4）在 src 目录下，创建一个 com.test.po 包，并在包中创建一个 User 类，该类用于封装 User 类型的请求参数。

```
package com.test.po;
public class User {
```

```java
private String username;
private Integer password;
public String getUsername() {
    return username;
}
public void setUsername(String username) {
    this.username = username;
}
public Integer getPassword() {
    return password;
}
public void setPassword(Integer password) {
    this.password = password;
}
@Override
public String toString() {
    return "User [username=" + username + ", password=" + password + "]";
}
}
```

(5)在 web>WEB-INF>jsp 下,创建页面文件 index.jsp 来测试 JSON 数据交互。

①导入 jquery 类库。

②使用 jquery 发送 AJAX 请求。

```html
<head>
<meta http-equiv="Content-Type" content="text/html; charset=UTF-8">
<title>Insert title here</title>
<script type="text/javascript"
src="${pageContext.request.contextPath}/js/jquery-1.11.3.min.js"></script>
<script type="text/javascript">
    function toTest() {
        var username=$("#username").val();
        var password=$("#password").val();
$.ajax({
            url:"${pageContext.request.contextPath}/test",
type:"post",
            data:Json.stringify(
{username:username,
password:password}),
dataType:"json",
success:function(data){
            if(data!=null){
                alert("您输入的用户名为:"+data.username+"密码为:"+data.password);
            }
        }
    });
}
```

```
</script>
</head>
<body>
<form>
用户名:<input type="text" name="username" id="username"><br/>
密码:<input type="password" name="password" id="password"><br/>
<input type="button" value="测试" onclick="toTest()"/>
</form>
</body>
```

将 jquery-1.11.3.min.js 放入 web 目录下 js 文件夹中,如图 12-2 所示。

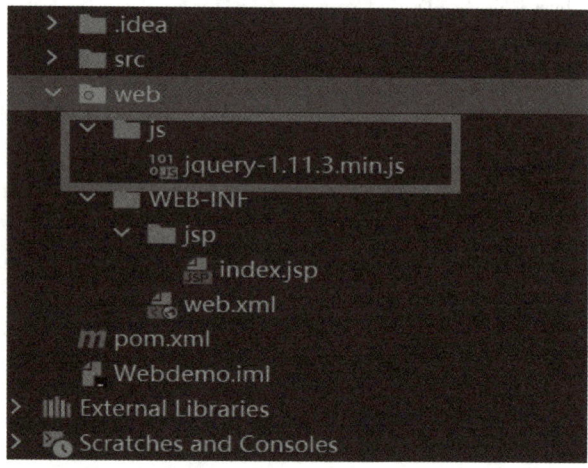

图 12-2　jquery-1.11.3.min.js 的引入

(6)在 src>main>java 目录下,创建一个 com.test.controller 包,在该包下创建一个用于用户操作的控制器类 UserController。

```
package com.test.controller;
import org.springframework.stereotype.Controller;
import org.springframework.web.bind.annotation.RequestBody;
import org.springframework.web.bind.annotation.RequestMapping;
import org.springframework.web.bind.annotation.ResponseBody;
import com.ex.po.User;
@Controller
public class UserController {
    @RequestMapping("/test")
    @ResponseBody
    public User test(@RequestBody User user) {
        System.out.println(user);
        return user;
    }
}
```

@RequestBody 用于将前端请求体中的 JSON 格式数据绑定到形参 user 上。

@ResponseBody 用于直接返回 User 对象(当返回 POJO 对象时,会默认转换成 JSON 格式数据进行响应)。

(7)将 demo 项目发布到 Tomcat 服务器并启动浏览器进行访问,如图 12-3 所示。

图 12-3 测试

2. 使用<bean>标签方式的 JSON 转换器配置

在配置 JSON 转换器时,除了常用的<mvc:annotation-driven>标签外,还可以使用<bean>标签的方式进行显式配置。

```
<bean class="org.springframework.web.servlet.mvc.method.annotation
.RequestMappingHandlerMapping"/>
<bean class="org.springframework.web.servlet.mvc.method.annotation
.RequestMappingHandlerAdapter">
    <property name="messageConverters">
        <list>
            <bean class="org.springframework.http.converter.json
.MappingJackson2HttpMessageConverter"/>
        </list>
    </property>
</bean>
```

3. 配置静态资源访问的方式

除了使用<mvc:resources>元素可以实现对静态资源的访问外,还有另外两种静态资源访问的配置方法。

(1)使用<mvc:defaule-servlet-handler>标签。

在 springmvc-config.xml 文件中,使用<mvc:defaule-servlet-handler>标签。

```
<mvc:default-servlet-handler/>
```

配置<mvc:defaule-servlet-handler>后,会在 Spring MVC 上下文中定义一个 org.springframework.web.servlet.resource.DefaultServletHttpRequestHandler(即默认的 Servlet 请求处理器)类型对象,它会像一个检察员,对进入 DispatcherServlet 的 URL 进行筛选。如果发现是静态资源的请求,就将该请求转由 Web 服务器默认的 Servlet 处理,默认的 Servlet 就会对这些资源放行;如果不是静态资源的请求,才由 DispatcherServlet 继续处理。

注意,一般 Web 服务器默认的 Servlet 名称是"default",因此 DefaultServletHttpRequest

Handler 可以找到它。如果使用的 Web 应用服务器默认的 Servlet 名称不是"default",则需要通过 default-servlet-name 属性显式指定,具体方式如下所示。

```
<mvc:default-servlet-handler default-servlet-name="Servlet 名称"/>
```

(2) 激活 Tomcat 默认的 Servlet 来处理静态文件访问。

激活 Tomcat 默认的 Servlet 时,需要在 web.xml 中添加以下内容。

```
<servlet-mapping>
<servlet-name>default</servlet-name>
<url-pattern>*.js</url-pattern>
</servlet-mapping>
<servlet-mapping>
<servlet-name>default</servlet-name>
<url-pattern>*.css</url-pattern>
</servlet-mapping>
```

上述代码中,配置了<servlet-mapping>来激活 Tomcat 默认的 Servlet 来处理静态文件,我们还可以根据需要继续追加<servlet-mapping>。此种方式和上一种方式本质上来说是一样的,都是使用 Web 服务器默认的 Servlet 来处理静态资源文件的访问。其中<servlet-name>中的值也是由使用的服务器来确定的,不同服务器需要使用不同的名称。

12.3 RESTful 支持

Spring MVC 除了支持 JSON 数据交换外,还支持 RESTful 风格的编程。

RESTful 也被称之为"REST(Representational State Transfer)",可以将它理解为一种软件架构风格或设计风格,而不是一个标准。

简单来说,RESTful 风格就是把请求参数变成请求路径的一种风格。

常规风格的请求如下所示。

```
http://.../queryItems?id=1
```

采用 RESTful 风格的请求如下所示。

```
http://.../Items/1
```

从上述两个请求中可以看出,RESTful 风格中的 URL 将请求参数"id=1"变成了请求路径的一部分,并且 URL 中的"queryItems"也变成了"Items"(RESTful 风格中的 URL 不存在动词形式的路径)。

RESTful 请求中,使用 put、delete、post、get 方式分别对应添加、删除、修改、查询的操作。

第12章 JSON 数据交互和 RESTful 支持

12.4 应用案例——用户信息查询

（1）在控制器类 UserController 中，编写用户查询方法 queryUser()。

```java
package com.test.controller;
import org.springframework.stereotype.Controller;
import org.springframework.web.bind.annotation.PathVariable;
import org.springframework.web.bind.annotation.RequestBody;
import org.springframework.web.bind.annotation.RequestMapping;
import org.springframework.web.bind.annotation.RequestMethod;
import org.springframework.web.bind.annotation.ResponseBody;
import com.ex.po.User;
@Controller
public class UserController {
    @RequestMapping("/test")
    @ResponseBody
    public User test(@RequestBody User user) {
        System.out.println(user);
        return user;
    }
    @RequestMapping(value = "/user/{id}", method = RequestMethod.GET)
    @ResponseBody
    public User queryUser(@PathVariable("id") String id) {
        System.out.println("id=" + id);
        User user = new User();
        if (id.equals("1")) {
            user.setUsername("peter");
        }
        return user;
    }
}
```

@RequestMapping(value = "/user/{id}", method = RequestMethod.GET)用于匹配请求路径（包括参数）和方法。其中 value = "/user/{id}" 表示可以匹配以"/user/{id}"结尾的请求，id 为请求中的动态参数；method = RequestMethod.GET 表示只接受 GET 方式的请求。方法中的 @PathVariable("id")用于接收并绑定请求参数，它可以将请求 URL 中的变量映射到方法的形参上。如果请求路径为"/user/{id}"，即请求参数中的 id 和方法形参名称一样，则 @PathVariable 后面的("id")可以省略。

（2）在 web>WEB-INF>jsp 目录下，编写页面文件 restful.jsp，在页面中使用 AJAX 方式通过输入的用户编号来查询用户信息。

```
<head>
<meta http-equiv="Content-Type" content="text/html; charset=UTF-8">
<title>Insert title here</title>
<script type="text/javascript"
    src="${pageContext.request.contextPath}/js/jquery-1.11.3.min.js"></script>
<script type="text/javascript">
    function test(){
        var id=$("#id").val();
        $.ajax({
            url:"${pageContext.request.contextPath}/user/"+id,
            type:"GET",
            dataType:"json",
            success:function(data){
                if(data!=null){
                    alert("您的用户名为:"+data.username);
                }
            }
        })
    }
</script>
</head>
<body>
<form>
编号:<input type="text" name="id" id="id">
<input type="button" value="测试" onclick="test()"/>
</form>
</body>
```

将该项目部署到 Tomcat 中测试即可看到弹框显示"peter"。

本章小结

  本章主要对 Spring MVC 中的 JSON 数据交互和 RESTful 风格的请求进行了详细的讲解。首先简单介绍了 JSON 的概念、作用和结构，然后通过案例讲解了 Spring MVC 中如何实现 JSON 数据的交互；接下来讲解了什么是 RESTful，最后通过用户信息查询案例来演示 RESTful 的实际使用。通过本章的学习，读者可以掌握 Spring MVC 中的 JSON 数据交互和对 RESTful 风格的支持，这对今后实际开发有极大的帮助。

第 12 章 JSON 数据交互和 RESTful 支持

课后习题

1. 请简述 JSON 数据交互两个注解的作用。
2. 请简述静态资源访问的几种配置方式。

第 13 章 拦截器

学习目标

[本章知识点]
　　拦截器的定义
　　拦截器的配置
　　单个拦截器的执行流程
　　多个拦截器的执行流程
　　应用案例——实现用户登录权限验证

[思政目标]
　　提高学生在沟通表达、自我学习和团队协作方面的能力；帮助学生在今后的职场中，在身处的软件开发团队中，能够与同事协同攻关、合作共赢。

13.1 拦截器概述

Spring MVC 中的拦截器(Interceptor)类似于 Servlet 中的过滤器(Filter)，它主要用于拦截用户请求并作相应的处理。例如通过拦截器可以进行权限验证、记录请求信息的日志、判断用户是否登录等。

13.1.1 拦截器的定义

要使用 Spring MVC 中的拦截器，就需要对拦截器类进行定义和配置。通常拦截器类可以通过两种方式来定义。

（1）通过实现 HandlerInterceptor 接口，或继承 HandlerInterceptor 接口的实现类（如 HandlerInterceptorAdapter）来定义。

（2）通过实现 WebRequestInterceptor 接口，或继承 WebRequestInterceptor 接口的实现类来定义。

以实现 HandlerInterceptor 接口方式为例，自定义拦截器类的代码如下所示。

```
public class CustomInterceptor implements HandlerInterceptor{
    public boolean preHandle(HttpServletRequest request, HttpServletResponse response, Object handler) throws Exception{
        return false;
    }
    public void postHandle(HttpServletRequest request, HttpServletResponse response, Object handler, ModelAndView modelAndView) throws Exception{
    }
    public void afterCompletion(HttpServletRequest request, HttpServletResponse response, Object handler, Exception ex) throws Exception{
    }
}
```

上述代码中，自定义拦截器实现了 HandlerInterceptor 接口，并实现了接口中的三个方法。

preHandle()方法：该方法会在控制器方法前执行，其返回值表示是否中断后续操作。当其返回值为 true 时，表示继续向下执行；当其返回值为 false 时，会中断后续的所有操作（包括调用下一个拦截器和控制器类中的方法执行等）。

postHandle()方法：该方法会在控制器方法调用之后、解析视图之前执行。可以通过此方法对请求域中的模型和视图做出进一步的修改。

afterCompletion()方法：该方法会在整个请求完成即视图渲染结束之后执行。可以通过此方法实现一些资源清理、记录日志信息等工作。

13.1.2 拦截器的配置

开发拦截器就像开发 Servlet 或者 Filter 一样，都需要在配置文件进行配置，配置代码如下所示。

```xml
<!--配置拦截器-->
<mvc:interceptors>
<!--<bean class="com.yyzy.interceptor.CustomeInterceptor" />-->
<!--拦截器1-->
<mvc:interceptor>
<!--配置拦截器的作用路径-->
<mvc:mapping path="/**"/>
<mvc:exclude-mapping path=""/>
<!--定义在<mvc:interceptor>下面的表示匹配指定路径的请求才进行拦截-->
<bean class="com.yyzy.interceptor.Intercptor1"/>
</mvc:interceptor>
```

```
<!--拦截器2-->
<mvc:interceptor>
<mvc:mapping path="/hello"/>
<bean class="com.yyzy.interceptor.Interceptor2"/>
</mvc:interceptor>
```

上面的代码中，<mvc:interceptors>元素用于配置一组拦截器，其子元素<bean>中定义的是全局拦截器，它会拦截所有的请求；而<mvc:interceptor>元素中定义的是指定路径的拦截器，它会对指定路径下的请求生效。<mvc:interceptor>元素的子元素<mvc:mapping>用于配置拦截器作用的路径，该路径在其属性 path 中定义。如上述代码中 path 的属性值"/**"表示拦截所有路径，"/hello"表示拦截所有以"/hello"结尾的路径。如果在请求路径中包含不需要拦截的内容，还可以通过<mvc:exclude-mapping>元素进行配置。

注意，<mvc:interceptor>中的子元素必须按照上述代码中的配置顺序进行编写，即<mvc:mapping><mvc:exclude-mapping><bean>，否则文件会报错。

13.2 拦截器的执行流程

13.2.1 单个拦截器的执行流程

在运行程序时，拦截器的执行是有一定顺序的，该顺序与配置文件中所定义的拦截器的顺序相关。单个拦截器在程序中的执行流程如图 13-1 所示。

（1）程序先执行 preHandle()方法，如果该方法的返回值为 true，则程序会继续向下执行处理器中的方法，否则将不再向下执行。

（2）在业务处理器（即控制器 Controller 类）处理完请求后，会执行 postHandle()方法，然后会通过 DispatcherServlet 向客户端返回响应。

（3）在 DispatcherServlet 处理完请求后，才会执行 afterCompletion()方法。

下面通过一个测试程序来验证它的执行流程。

新建一个 Web 项目，准备好 Spring MVC 程序运行所需要的 jar 包，在 web.xml 中配置前端过滤器和初始化加载信息。

新建一个测试 Controller，代码如下所示。

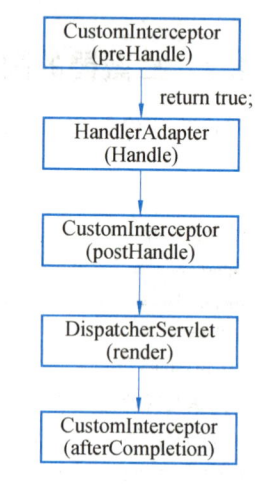

图 13-1 单个拦截器的执行流程

第13章 拦截器

```
@Controller
public class HelloController{
@RequestMapping("/hello")
public String Hello(){
System.out.println("Hello!");
return "success";
}
}
```

然后，新建一个拦截器，实现 HandlerInterceptor 接口，并实现其中的方法。

```
public class CustomeInterceptor implements HandlerInterceptor{
@Override
public booleanpreHandle(HttpServletRequest httpServletRequest,HttpServletResponse httpServletResponse,Object o)
throws Exception{
System.out.println("CustomInterceptor....preHandle");
//对浏览器的请求进行放行处理
return true;
}
@Override
public void postHandle(HttpServletRequest httpServletRequest,HttpServletResponse httpServletResponse,Object o,ModelAndView modelAndView)
throws Exception{
System.out.println("CustomInterceptor....postHandle");
}
@Override
public void afterCompletion(HttpServletRequest httpServletRequest,HttpServletResponse httpServletResponse,Object o,Exception e)
throws Exception{
System.out.println("CustomInterceptor....afterCompletion");
}
}
```

在配置文件中配置拦截器。

```
<beans xmlns="http://www.springframework.org/schema/beans" xmlns:xsi="http://www.w3.org/2001/XMLSchema-instance" xmlns:mvc="http://www.springframework.org/schema/mvc" xmlns:context="http://www.springframework.org/schema/context" xsi:schemaLocation="http://www.springframework.org/schema/beans http://www.springframework.org/schema/beans/spring-beans-4.3.xsd http://www.springframework.org/schema/mvc http://www.springframework.org/schema/mvc/spring-mvc-4.3.xsd http://www.springframework.org/schema/context http://www.springframework.org/schema/context/spring-context-4.3.xsd">
<!--定义组件扫描器,指定需要扫描的包-->
<context:component-scan base-package="com.yyzy.controller"/>
<!--配置视图解析器 -->
<bean class="org.springframework.web.servlet.view.InternalResourceViewResolver">
<property name="prefix" value="/WEB-INF/jsp/"/>
<property name="suffix" value=".jsp"/>
</bean>
<!--配置拦截器-->
```

```
<mvc:interceptors><bean class="com.yyzy.interceptor.CustomeInterceptor" />
</beans>
```

程序先执行拦截器类中的 preHandle() 方法，然后执行控制器中的 Hello() 方法，最后分别执行拦截器类中的 postHandle() 方法和 afterCompletion() 方法。这与上文所描述的单个拦截器的执行顺序是一致的。

13.2.2 多个拦截器的执行流程

多个拦截器（假设有两个拦截器 Interceptor1 和 Interceptor2，并且在配置文件中，Interceptor1 拦截器配置在前）在程序中的执行流程如图 13-2 所示。

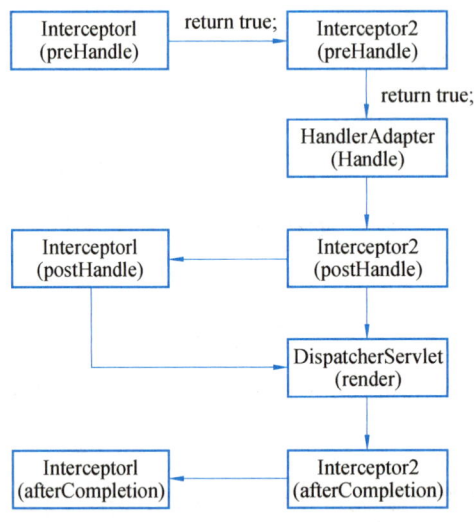

图 13-2 多个拦截器的执行流程

当有多个拦截器同时工作时，它们的 preHandle() 方法会按照配置文件中拦截器的配置顺序执行，而它们的 postHandle() 方法和 afterCompletion() 方法则会按照配置顺序的反序执行。

下面是测试案例。

新建两个拦截器。

```
//第一个拦截器
public class Intercptor1 implements HandlerInterceptor {
    @Override
    public booleanpreHandle(HttpServletRequest httpServletRequest, HttpServletResponse httpServletResponse, Object o) throws Exception { System.out.println("Interceptor1....preHandle");
        return true;
    }
    @Override
    public void postHandle(HttpServletRequest httpServletRequest, HttpServletResponse httpServletResponse, Object o, ModelAndViewmodelAndView) throws Exception { System.out.println("Interceptor1....postHandle");
```

```
    }
    @Override
    public void afterCompletion(HttpServletRequest httpServletRequest, HttpServletResponse httpServletResponse, Object o, Exception e) throws Exception {
        System.out.println("Interceptor1....afterCompletion");
    }
}
//第二个拦截器
public class Interceptor2 implements HandlerInterceptor {
    @Override
    public boolean preHandle(HttpServletRequest httpServletRequest, HttpServletResponse httpServletResponse, Object o) throws Exception {
        System.out.println("Interceptor2....preHandle");
        return true;
    }
    @Override
    public void postHandle(HttpServletRequest httpServletRequest, HttpServletResponse httpServletResponse, Object o, ModelAndView modelAndView) throws Exception {
        System.out.println("Interceptor2....postHandle");
    }
    @Override
    public void afterCompletion(HttpServletRequest httpServletRequest, HttpServletResponse httpServletResponse, Object o, Exception e) throws Exception {
        System.out.println("Interceptor2....afterCompletion");
    }
}
```

添加配置信息。

```
<!--拦截器1-->
<mvc:interceptor>
<!--配置拦截器的作用路径-->
<mvc:mapping path="/**"/>
<!--定义在<mvc:interceptor>下面的表示匹配指定路径的请求才进行拦截-->
<bean class="com.yyzy.interceptor.Intercrptor1"/></mvc:interceptor>
<!--拦截器2-->
<mvc:interceptor><mvc:mapping path="/hello"/>
<bean class="com.yyzy.interceptor.Interceptor2"/>
</mvc:interceptor>
```

从上面的代码可以看出，程序先执行前两个拦截器类中的 preHandle()方法，这两个方法的执行顺序与配置文件中定义的顺序相同；然后执行控制器类中的 Hello()方法；最后执行两个拦截器类中的 postHandle()方法和 afterCompletion()方法，且这两个方法的执行顺序与配置文件中所定义的拦截器顺序相反。

13.3 应用案例——用户认证拦截

用户访问系统的资源(URL)，如果用户没有进行身份认证，系统将进行拦截，并跳转登录页面；如果用户已经认证通过，用户可以继续访问系统的资源。

拦截实现步骤如下。

(1)使用 Maven 新建项目，选择模板 maven-archetype-webapp，搭建 Spring MVC 项目。

（2）在 pom.xml 中添加依赖 spring-context、spring-web、spring-webmvc。Spring-context 是 Spring 框架的依赖包，Spring MVC 框架需要引入两个依赖包：spring-web 和 spring-mvc。还会应用到 HttpServletRequest 等类，需要引入 Servlet 的 jar 包，所以还需要添加 servlet-api 的依赖。

```xml
<dependencies>
<dependency>
<groupId>org.springframework</groupId>
<artifactId>spring-context</artifactId>
<version>5.1.7.RELEASE</version>
</dependency>
<dependency>
<groupId>org.springframework</groupId>
<artifactId>spring-web</artifactId>
<version>5.1.7.RELEASE</version>
</dependency>
<dependency>
<groupId>org.springframework</groupId>
<artifactId>spring-webmvc</artifactId>
<version>5.1.7.RELEASE</version>
</dependency>
<dependency>
<groupId>javax.servlet</groupId>
<artifactId>servlet-api</artifactId>
<version>2.5</version>
</dependency>
</dependencies>
```

（3）在 web.xml 文件中配置 DispatcherServlet。

```xml
<web-app version="2.4"
xmlns="http://java.sun.com/xml/ns/j2ee"
xmlns:xsi="http://www.w3.org/2001/XMLSchema-instance"
xsi:schemaLocation="http://java.sun.com/xml/ns/j2ee http://java.sun.com/xml/ns/j2ee/web-app_2_4.xsd">
<servlet>
<display-name>DispatcherServlet</display-name>
<servlet-name>springMVC</servlet-name>
<servlet-class>org.springframework.web.servlet.DispatcherServlet</servlet-class>
</servlet>
<servlet-mapping>
<servlet-name>springMVC</servlet-name>
<url-pattern>/</url-pattern>
</servlet-mapping>
</web-app>
```

（4）在 Java 下面创建包 com.yyzy.po，在该包中创建实体类 User 类。

```java
public class User{
    int id;
    String username;
    String password;
    //get/set 方法省略
}
```

(5)在 webapp 下的 WEB-INF 下创建 login.jsp 页面。

```jsp
<body>
${msg}
<form action="login" method="post">
用户名:<input type="text" name="username"><br>
密码:<input type="password" name="password"><br>
<input type="submit" value="登录">
</form>
</body>
```

(6)创建 Controller 类。

在 Java 文件夹下创建 com.yyzy.controller 包,在该包中创建控制器类 UserController,该类需要实现 Controller 接口。

```java
@Controller
public class UserController{
//跳转到登录页面
    @RequestMapping(value="/login")
    public String toLogin(){
        return "login";
    }
    //用户登录
@RequestMapping(value="/login2")
    public String userLogin(User user, Model model, HttpSession session){
        //获取用户名和密码
String username=user.getUsername();
        String pass=user.getPassword();
        //获取用户名和密码后进行判断
if(username!=null&&username.equals("zhangsan")&&pass!=null&&pass.equals("123456")){
            //将对象添加到 session 中
session.setAttribute("u",user);
            return "main";
        }
        model.addAttribute("msg","用户名或密码错误,请重新登录!");
            return "login";
    }
//退出登录
    @RequestMapping(value="/logout")
    public String logout(HttpSession session){
```

```
            //清除 Session
    session.invalidate();
            return "login";
    }
```

(7) 在 webapp 下的 WEB-INF 下创建 main.jsp 页面。

```
<body>
当前用户:${u}<br>
<a href="logout">退出</a>
</body>
```

(8) 在 Java 下创建包 com.yyzy.interceptor，在包中创建拦截器 LoginInterceptor。

```
public class LoginInterceptor implements HandlerInterceptor {
    public boolean preHandle(HttpServletRequest request, HttpServletResponse response, Object handler) throws Exception {
        //获取请求的 URL
        String url=request.getRequestURI();
        //除了 login.jsp 是可以公开访问的,其他的 URL 都进行拦截
        if(url.indexOf("login")>=0){
            return true;
        }
        //获取 Session
        HttpSession session=request.getSession();
        User user=(User)session.getAttribute("u");
        //判断 Session 中是否有用户数据,若有,则返回 true,继续向下执行
        if(user!=null){
            return true;
        }
        //不符合条件的转发到登录页面
        request.setAttribute("msg","您没有登录,请先登录");
        request.getRequestDispatcher("WEB-INF/jsp/login.jsp").forward(request,response);
        return false;
    }
    public void postHandle(HttpServletRequest request, HttpServletResponse response, Object handler, ModelAndView modelAndView) throws Exception {
    }
    public void afterCompletion(HttpServletRequest request, HttpServletResponse response, Object handler, Exception ex) throws Exception {
    }
}
```

在上面程序中的 preHandle() 方法中，先获取请求的 URL，然后通过 indexOf() 方法判断 URL 中是否有"/login"字符串。如果有，则返回 true，即直接放行；如果没有，则继续向下执行拦截处理。接下来获取 Session 中的用户信息，如果 Session 中包含用户信息，即表示用户已登录，也直接放行；否则会转发到登录页面，不再执行后续程序。

第 13 章　拦截器

（9）创建 Spring 框架配置文件。在 WEB-INF 下面创建 Spring MVC 框架配置文件 springMVC-servlet.xml。

```
<?xml version="1.0" encoding="UTF-8"?>
<beans xmlns="http://www.springframework.org/schema/beans"
xmlns:mvc="http://www.springframework.org/schema/mvc"
xmlns:xsi="http://www.w3.org/2001/XMLSchema-instance"
xmlns:context="http://www.springframework.org/schema/context"
xsi:schemaLocation="http://www.springframework.org/schema/beans
        http://www.springframework.org/schema/beans/spring-beans-4.3.xsd
        http://www.springframework.org/schema/mvc
        http://www.springframework.org/schema/mvc/spring-mvc-4.3.xsd
        http://www.springframework.org/schema/context
        http://www.springframework.org/schema/context/spring-context-4.3.xsd">
<context:component-scan base-package="com.yyzy.controller"></context:component-scan>
<!--配置视图解析器-->
<bean class="org.springframework.web.servlet.view.InternalResourceViewResolver">
<property name="prefix" value="/WEB-INF/jsp/"></property>
<property name="suffix" value=".jsp"></property>
</bean>
<mvc:interceptors>
<mvc:interceptor>
<mvc:mapping path="/**"/>
<bean class="com.yyzy.interceptor.LoginInterceptor"></bean>
</mvc:interceptor>
</mvc:interceptors>
</beans>
```

注意，在 web.xml 中 servlet 中如果没有配置 <init-param> 子元素，则应用程序会默认到 WEB-INF 下寻找如下方式命名的配置文件。

　　servletName-servlet.xml

（10）测试。配置 Tomcat 服务器，在浏览器中输入"http://localhost：8080/chapter15/login"，查看是否正常显示。

本章小结

　　本章主要讲述了拦截器的定义和配置，以及单个拦截器和多个拦截器的执行流程，并通过一个案例来说明。

课后习题

1. 请简述 Spring MVC 拦截器的定义方式。
2. 请简述单个拦截器和多个拦截器的执行流程。

第 14 章
SSM 框架整合

 学习目标

[本章知识点]
  框架的整合思路
  整合所需 jar 包介绍
  编写配置文件
  整合应用测试

[思政目标]
  引导学生提升专业认同感，稳定专业心态，学习好本专业的知识，掌握好本专业的技能。

  通过前面章节的学习，相信读者已经掌握了 Spring、MyBatis、Spring MVC 框架的使用，并且已经掌握了 Spring 与 MyBatis 的整合使用。在实际项目开发中，这三大框架通常都会整合在一起使用。本章就对 SSM（Spring、Spring MVC、MyBatis）框架的整合使用进行详细的讲解。

  在第 10 章讲解 Spring 与 MyBatis 框架的整合时，我们是通过 Spring 实例化 Bean，然后调用实例对象中的查询方法来执行 MyBatis 映射文件中的 SQL 语句，如果能够正确查询出数据库中的数据，那么我们就认为 Spring 与 MyBatis 框架整合成功。同样，在学习完 Spring MVC 后，如果我们可以通过前台页面来执行查询方法，并且查询出的数据能够在页面中正确显示，那么我们也可以认为三大框架整合成功。

14.1　SSM 整合 jar 包介绍

要实现 SSM 框架的整合，首先要准备这三个框架的 jar 包，以及其他整合所需的 jar 包。在第 10 章讲解 Spring 与 MyBatis 框架整合时，已经介绍了 Spring 与 MyBatis 整合所需要的 jar 包，这里只需要再加入 Spring MVC 的相关 jar 包即可。具体如下：

- spring-web-4.0.2.RELEASE.jar
- spring-webmvc-4.0.2.RELEASE.jar

需要返回 JSON 数据的话加入 JSON 的 jar 包，具体如下：

- jackson-core.2.9.3.jar
- jackson-databind.2.9.3.jar
- jackson-annotations.2.9.3.jar

对 SSM 整合时所需的全部 jar 包依赖注入，如下所示。

```xml
<properties>
<maven.compiler.source>8</maven.compiler.source>
<maven.compiler.target>8</maven.compiler.target>
<project.build.sourceEncoding>UTF-8</project.build.sourceEncoding>
<spring.version>4.0.2.RELEASE</spring.version>
<mybatis.version>3.2.8</mybatis.version>
<slf4j.version>1.7.12</slf4j.version>
<log4j.version>1.2.17</log4j.version>
<jackson.version>2.9.3</jackson.version>
</properties>
<dependencies>
<!-- Spring 框架包 start -->
<dependency>
<groupId>org.springframework</groupId>
<artifactId>spring-test</artifactId>
<version>${spring.version}</version>
</dependency>
<dependency>
<groupId>org.springframework</groupId>
<artifactId>spring-core</artifactId>
<version>${spring.version}</version>
</dependency>
<dependency>
<groupId>org.springframework</groupId>
<artifactId>spring-oxm</artifactId>
<version>${spring.version}</version>
</dependency>
<dependency>
```

```xml
<groupId>org.springframework</groupId>
<artifactId>spring-tx</artifactId>
<version>${spring.version}</version>
</dependency>
<dependency>
<groupId>org.springframework</groupId>
<artifactId>spring-jdbc</artifactId>
<version>${spring.version}</version>
</dependency>
<dependency>
<groupId>org.springframework</groupId>
<artifactId>spring-aop</artifactId>
<version>${spring.version}</version>
</dependency>
<dependency>
<groupId>org.springframework</groupId>
<artifactId>spring-context</artifactId>
<version>${spring.version}</version>
</dependency>
<dependency>
<groupId>org.springframework</groupId>
<artifactId>spring-context-support</artifactId>
<version>${spring.version}</version>
</dependency>
<dependency>
<groupId>org.springframework</groupId>
<artifactId>spring-expression</artifactId>
<version>${spring.version}</version>
</dependency>
<dependency>
<groupId>org.springframework</groupId>
<artifactId>spring-orm</artifactId>
<version>${spring.version}</version>
</dependency>
<!-- Spring 框架包 end -->
<!--MyBatis 框架包 start -->
<dependency>
<groupId>org.mybatis</groupId>
<artifactId>mybatis</artifactId>
<version>${mybatis.version}</version>
</dependency>
<dependency>
<groupId>org.mybatis</groupId>
<artifactId>mybatis-spring</artifactId>
<version>1.2.2</version>
</dependency>
<dependency>
<groupId>org.springframework</groupId>
```

```xml
    <artifactId>spring-web</artifactId>
    <version>${spring.version}</version>
</dependency>
<dependency>
    <groupId>org.springframework</groupId>
    <artifactId>spring-webmvc</artifactId>
    <version>${spring.version}</version>
</dependency>
<!--MyBatis 框架包 end -->
<!--JSON -->
<dependency>
    <groupId>com.fasterxml.jackson.core</groupId>
    <artifactId>jackson-core</artifactId>
    <version>${jackson.version}</version>
</dependency>
<dependency>
    <groupId>com.fasterxml.jackson.core</groupId>
    <artifactId>jackson-databind</artifactId>
    <version>${jackson.version}</version>
</dependency>
<dependency>
    <groupId>com.fasterxml.jackson.core</groupId>
    <artifactId>jackson-annotations</artifactId>
    <version>${jackson.version}</version>
</dependency>
<!--JSON -->
<!--数据库驱动 -->
<dependency>
    <groupId>mysql</groupId>
    <artifactId>mysql-connector-java</artifactId>
    <version>5.1.35</version>
</dependency>
<!--数据库驱动-->
<!-- log start -->
<dependency>
    <groupId>log4j</groupId>
    <artifactId>log4j</artifactId>
    <version>${log4j.version}</version>
</dependency>
<dependency>
    <groupId>org.slf4j</groupId>
    <artifactId>slf4j-api</artifactId>
    <version>${slf4j.version}</version>
</dependency>
<dependency>
    <groupId>org.slf4j</groupId>
    <artifactId>slf4j-log4j12</artifactId>
    <version>${slf4j.version}</version>
```

```
        </dependency>
        <!-- log END -->
</dependencies>
```

14.2 整合步骤

项目结构如图14-1所示。

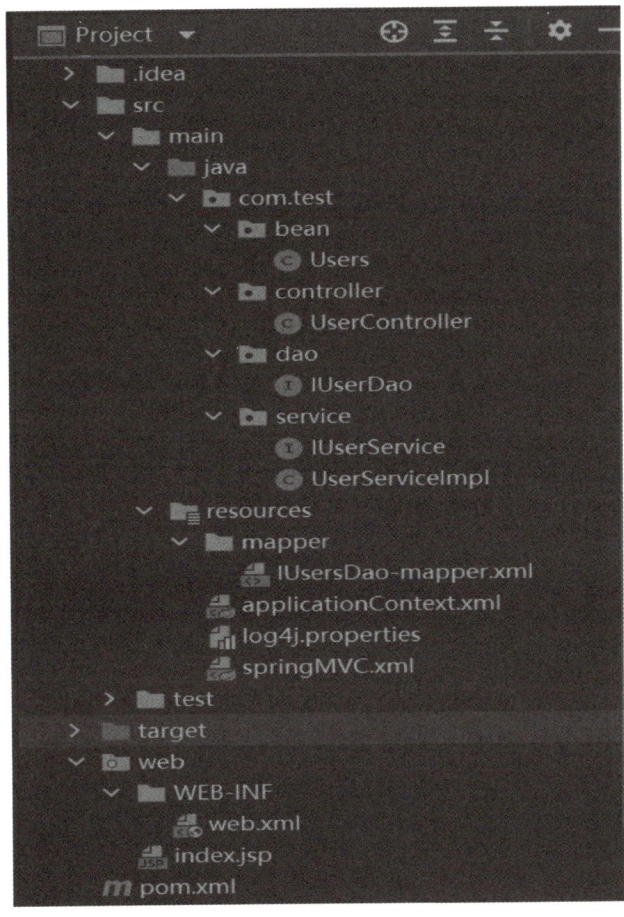

图14-1 项目结构图

(1)创建 Maven(Web)项目。
(2)在 com.test.bean 下创建实体类(Users.java)。

```
package com.test.bean;

public class Users {
    private Integer id;
```

```java
    private String name;
    private String pwd;

    public Users() {
        super();
        // TODO Auto-generated constructor stub
    }
    public Users(Integer id, String name, String pwd) {
        super();
        this.id = id;
        this.name = name;
        this.pwd = pwd;
    }
    public Integer getId() {
        return id;
    }
    public void setId(Integer id) {
        this.id = id;
    }
    public String getName() {
        return name;
    }
    public void setName(String name) {
        this.name = name;
    }
    public String getPwd() {
        return pwd;
    }
    public void setPwd(String pwd) {
        this.pwd = pwd;
    }

}
```

（3）创建 Users 的数据访问接口（IUserDao.java）。

```java
package com.test.dao;
import java.util.List;
import com.test.bean.Users;
public interface IUserDao{
    /**添加用户信息 */
    public Integer insertUser(Users users) throws Exception;
    /** 根据主键删除用户信息 */
    public Integer delUserByUid(Integer uid) throws Exception;
    /**根据主键修改用户信息 */
    public Integer updateUserByUid(Users users) throws Exception;
    /**根据主键查询用户信息*/
```

```
public Users getUserByUid(Integer uid) throws Exception;
/**查询所有用户信息*/
public List<Users> getUserList() throws Exception;
/**查询用户总数据量*/
public long getUserCount() throws Exception;
}
```

(4)在 resources 下创建 mapper 文件夹,并且在 mapper 文件下创建数据访问层映射文件(IUsersDao-mapper.xml)。

```xml
<!DOCTYPE mapper PUBLIC
"-//mybatis.org//DTD Mapper 3.0//EN"
    "http://mybatis.org/dtd/mybatis-3-mapper.dtd">
<mapper namespace="com.test.dao.IUserDao">
<insert id="insertUser" parameterType="com.test.bean.Users">
        insert into users(name,pwd) values (#{name},#{pwd})
</insert>
<delete id="delUserByUid" parameterType="int">
        delete from users where id=#{id}
</delete>
<update id="updateUserByUid" parameterType="com.test.bean.Users">
    update users set name=#{name},pwd=#{pwd} where id=#{id}
</update>
<select id="getUserByUid" parameterType="int" resultType="com.test.bean.Users">
        select * from users where id=#{id}
</select>
<select id="getUserList" resultType="com.test.bean.Users">
        select * from users
</select>
<select id="getUserCount" resultType="long">
        select count(*) from users
</select>
</mapper>
```

(5)在 resources 下编写 IoC 配置文件(applicationContext.xml)。

```xml
<beans xmlns="http://www.springframework.org/schema/beans"
xmlns:xsi="http://www.w3.org/2001/XMLSchema-instance"
xmlns:context="http://www.springframework.org/schema/context"
xsi:schemaLocation="http://www.springframework.org/schema/beans
http://www.springframework.org/schema/beans/spring-beans-4.0.xsd
    http://www.springframework.org/schema/context
    http://www.springframework.org/schema/context/spring-context-4.0.xsd">
<context:component-scan base-package="com.test.service"></context:component-scan>
<!--1.配置数据源 DriverManagerDataSource   -->
<bean id="ds" class="org.springframework.jdbc.datasource.DriverManagerDataSource">
<property name="driverClassName" value="com.mysql.jdbc.Driver"></property>
```

```xml
<property name="url" value="jdbc:mysql://localhost:3306/mybatisdb"></property>
<property name="username" value="root"></property>
<property name="password" value="root"></property>
</bean>
<!--2.配置 SqlSessionFactoryBean
dataSource:数据源
mapperLocations:mybatis 的映射文件
-->
<bean id="ssf" class="org.mybatis.spring.SqlSessionFactoryBean">
<property name="dataSource" ref="ds"></property>
<property name="mapperLocations" value="classpath:mapper/*-mapper.xml"></property>
</bean>
<!--3.为数据访问层接口配置一个扫描器 MapperScannerConfigurer
去扫描接口并且以动态代理的形式为接口生成实例对象
sqlSessionFactory:依赖于 SqlSessionFactoryBean 的实例
basePackage:为哪个包结构下的接口生成动态代理对象-->
<bean id="msc" class="org.mybatis.spring.mapper.MapperScannerConfigurer">
<!--<property name="sqlSessionFactory" ref="ssf"></property>-->
<property name="sqlSessionFactoryBeanName" value="ssf"></property>
<property name="basePackage" value="com.test.dao"></property>
</bean>
</beans>
```

在上面的代码中,首先配置数据源用于访问数据库,然后配置 SqlSessionFactoryBean 用于整合 MyBatis 框架的 MyBatis 工厂信息,接着定义 mapper 扫描器来扫描数据访问层接口,最后配置注解扫描器扫描 Service 层装配 servicebean。

在实际开发时,为了避免 Spring 配置文件中的信息过于臃肿,通常会将 Spring 配置文件中的信息按照不同的功能分散在多个配置文件中。例如,可以将事务配置放置在名称为"applicationContext-transaction.xml"的文件中,将数据源等信息放置在名称为"applicationContext-db.xml"的文件中等。这样,在 web.xml 中配置加载 Spring 文件信息时,只需通过 applicationContext-*.xml 的方式即可自动加载全部配置文件。

(6)在 com.test.service 下创建服务层接口(IUserService.java)以及实现类(UserServiceImpl.java)。

```java
package com.test.service;
import com.test.bean.Users;
import java.util.List;
public interface IUserService{
List<Users>getAllUserList() throws Exception;
}
package com.test.service;
import com.test.bean.Users;
import com.test.dao.IUserDao;
import org.springframework.beans.factory.annotation.Autowired;
```

```
import org.springframework.stereotype.Service;
import java.util.List;
@Service
public class UserServiceImplimplements IUserService{
@Autowired
private IUserDaodao;
@Override
public List<Users>getAllUserList() throws Exception{
return dao.getUserList();
    }
}
```

（7）在 web>WEB-INF 下配置 web.xml，用于加载 IoC 容器以及 MVC 容器。

```
<?xml version="1.0" encoding="UTF-8"?>
<web-app xmlns="http://xmlns.jcp.org/xml/ns/javaee"
xmlns:xsi="http://www.w3.org/2001/XMLSchema-instance"
xsi:schemaLocation="http://xmlns.jcp.org/xml/ns/javaee
http://xmlns.jcp.org/xml/ns/javaee/web-app_4_0.xsd"
version="4.0">
<!--配置监听器  用于加载 IoC 容器-->
<listener>
<listener-class>org.springframework.web.context.ContextLoaderListener</listener-class>
</listener>
<!--指定 IoC 容器位置-->
<context-param>
<param-name>contextConfigLocation</param-name>
<param-value>classpath:applicationContext.xml</param-value>
</context-param>
<!--配置 Spring MVC 中央控制器-->
<servlet>
<servlet-name>mvc</servlet-name>
<servlet-class>org.springframework.web.servlet.DispatcherServlet</servlet-class>
<!--指定 Spring MVC 容器位置-->
<init-param>
<param-name>contextConfigLocation</param-name>
<param-value>classpath:springMVC.xml</param-value>
</init-param>
</servlet>
<!--配置中央控制器处理请求格式-->
<servlet-mapping>
<servlet-name>mvc</servlet-name>
<url-pattern>*.do</url-pattern>
</servlet-mapping>
</web-app>
```

在上面的代码中，通过监听器加载 IoC 容器，配置 Spring MVC 的中央控制器并且加载 MVC 容器。

(7)在 resources 下编写 MVC 配置文件(springMVC.xml)。

```xml
<beans xmlns="http://www.springframework.org/schema/beans"
xmlns:xsi="http://www.w3.org/2001/XMLSchema-instance"
xmlns:context="http://www.springframework.org/schema/context"
xmlns:mvc="http://www.springframework.org/schema/mvc"
xsi:schemaLocation=" http://www.springframework.org/schema/beans  http://www.springframework.org/schema/beans/spring-beans-4.0.xsd
    http://www.springframework.org/schema/context
http://www.springframework.org/schema/context/spring-context-4.0.xsd
    http://www.springframework.org/schema/mvc
http://www.springframework.org/schema/mvc/spring-mvc-4.0.xsd">
<context:component-scan base-package="com.test.controller"></context:component-scan>
<mvc:annotation-driven/>
<bean class="org.springframework.web.servlet.view.InternalResourceViewResolver">
<property name="prefix" value="/views/" />
<property name="suffix" value=".jsp" />
</bean>
</beans>
```

在上面的代码中,主要配置用于扫描@Controller 注解的包扫描器、注解驱动器以及视图解析器。

(8)在 web 下创建 index.jsp,向后端发送请求。

```jsp
<%@ page contentType="text/html;charset=UTF-8" language="java" %>
<html>
<head>
<title>$Title$</title>
</head>
<body>
<a href="http://localhost:8080/SpringMyBatisDemo_war_exploded2/getUserList.do">获得用户数据</a>
</body>
</html>
```

(9)在 com.test.controller 下创建控制类(UserController.java)。

```java
package com.test.controller;
import com.test.bean.Users;
import com.test.service.IUserService;
import org.springframework.beans.factory.annotation.Autowired;
import org.springframework.stereotype.Controller;
import org.springframework.web.bind.annotation.RequestMapping;
import org.springframework.web.bind.annotation.ResponseBody;
import org.springframework.web.servlet.ModelAndView;
import java.util.List;
@Controller
public class UserController{
@Autowired
private IUserService userService;
```

```
@RequestMapping("/getUserList")
@ResponseBody
public List<Users>getUserList(){
ModelAndView mav=new ModelAndView();
try{
List<Users> list=userService.getAllUserList();
return list;
        }catch(Exception e){
e.printStackTrace();
        }
return null;
    }
}
```

（10）启动项目访问 index.jsp，单击超链接发送请求测试，出现如图14-2所示数据则表示 SSM 整合成功。

图14-2　测试

日志文件 log4j.properties 内容如下所示（不需要则可不添加，不影响 SSM 的整合）。

```
log4j.rootLogger=INFO,stdout
# SqlMap logging configuration...
log4j.logger.com.ibatis=DEBUG
log4j.logger.com.ibatis.common.jdbc.SimpleDataSource=DEBUG
log4j.logger.com.ibatis.sqlmap.engine.cache.CacheModel=DEBUG
log4j.logger.com.ibatis.sqlmap.engine.impl.SqlMapClientImpl=DEBUG
log4j.logger.com.ibatis.sqlmap.engine.builder.xml.SqlMapParser=DEBUG
log4j.logger.com.ibatis.common.util.StopWatch=DEBUG
log4j.logger.java.sql.Connection=DEBUG
log4j.logger.java.sql.Statement=DEBUG
log4j.logger.java.sql.PreparedStatement=DEBUG
log4j.logger.java.sql.ResultSet=DEBUG
# Console output...
log4j.appender.stdout=org.apache.log4j.ConsoleAppender
log4j.appender.stdout.layout=org.apache.log4j.PatternLayout
log4j.appender.stdout.layout.ConversionPattern=%5p [%t] - %m%n
```

本章小结

本章主要讲解了 SSM 框架的整合知识。首先对 SSM 框架整合的环境搭建进行了讲解，然后通过一个查询用户信息的案例讲解了具体的整合过程。通过本章的学习，读者能够了解 SSM 框架的整合思路，掌握 SSM 框架的整合过程以及如何使用。框架的整合是 SSM 框架使用的基础，读者一定要多加练习并熟练掌握。

课后习题

1. 请简述 SSM 框架整合思路。
2. 请简述 SSM 框架整合时，Spring 配置文件中的配置信息（无须写代码，简单描述所要配置的内容即可）。

第 15 章 教务管理系统

学习目标

[本章知识点]
  系统概述
  数据库设计
  项目结构
  jar 包依赖
  准备项目环境

[思政目标]
  引导学生利用新技术、新应用创新媒体传播方式，引导学生关注最新的传播技术、传播方式。

15.2 系统概述

本章将演示利用前面所学的 SSM 框架来实现一个简易的教务管理系统。该系统后台使用 SSM 框架编写，前台页面使用 jQuery 框架完成页面信息展示功能。系统中主要实现了如图 15-1 所示功能模块。

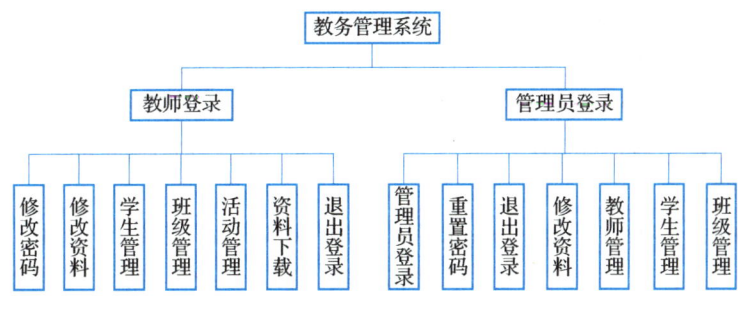

图 15-1 功能模块图

需求详情如下。

> **教务管理系统**
>
> 针对对象：
> 1. 系统管理员
> 2. 班主任
>
> 系统管理员功能需求：
> 1. 管理员登录。
> 2. 管理员重置密码。
> 3. 退出登录。
> 4. 修改本人资料。
> 5. 教师管理[分页展示老师信息，添加老师信息，开除老师(不删除仅修改状态)，禁止登录]。
> 6. 学生管理(分页展示学生信息，根据班级、学号、姓名模糊查询学生)。
> 7. 班级管理(分页展示班级信息开设班级，修改班级班主任)。
>
> 班主任功能需求：
> 1. 班主任登录。
> 2. 班主任修改密码。
> 3. 班主任修改个人资料。
> 4. 退出登录。
> 5. 学生管理[分页查询所带班级学生信息(根据姓名、学号、班级、状态模糊查询)，添加学生，修改学生信息，开除学生(不删除仅修改状态)]。
> 6. 班级管理(分页展示所带班级，单击班级在下方分页显示班级学生信息，修改班级资料，上传班级资料)。
> 7. 活动管理(创建活动，修改活动，分页展示活动，单击活动显示活动详情)。
> 8. 班级资料下载(资料展示，资料下载)。

15.2 系统架构设计

根据功能的不同，本系统项目结构可以划分为以下几个层次。

- 持久对象层(也称"持久层"或"持久化层")：该层由若干持久化类(实体类)组成。
- 数据访问层(DAO 层)：该层由若干 DAO 接口和 MyBatis 映射文件组成。接口的名称统一以 Dao 结尾，且 MyBatis 的映射文件名称要与接口的名称相同。
- 业务逻辑层(Service 层)：该层由若干 Service 接口和实现类组成。在本系统中，业务逻辑层的接口统一使用 Service 结尾，其实现类名称统一在接口名后加 Impl。该层主要用于实现系统的业务逻辑。
- Web 表现层：该层主要包括 Spring MVC 中的 Controller 类和 JSP 页面。Controller 类主要负责拦截用户请求，并调用业务逻辑层中相应组件的业务逻辑方法来处理用户请求，然后将相应的结果返回给 JSP 页面。

为了让读者更清晰地了解各个层次之间的关系，下面通过一张图来描述各个层次的关系和作用，如图 15-2 所示。

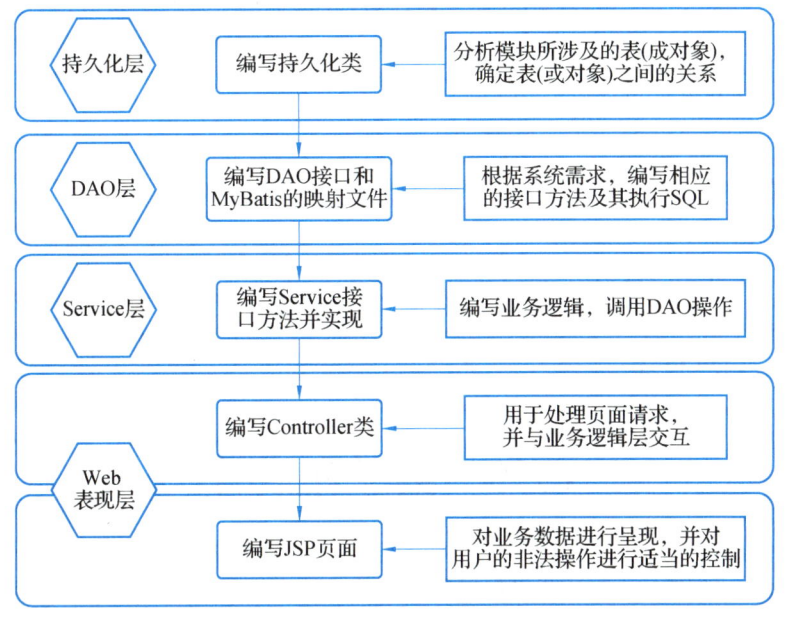

图 15-2　系统层次结构

15.3　文件组织结构

在正式讲解项目的编写之前，先来了解项目中所涉及的包文件、配置文件、页面文件等在项目中的组织结构，如图 15-3 所示。

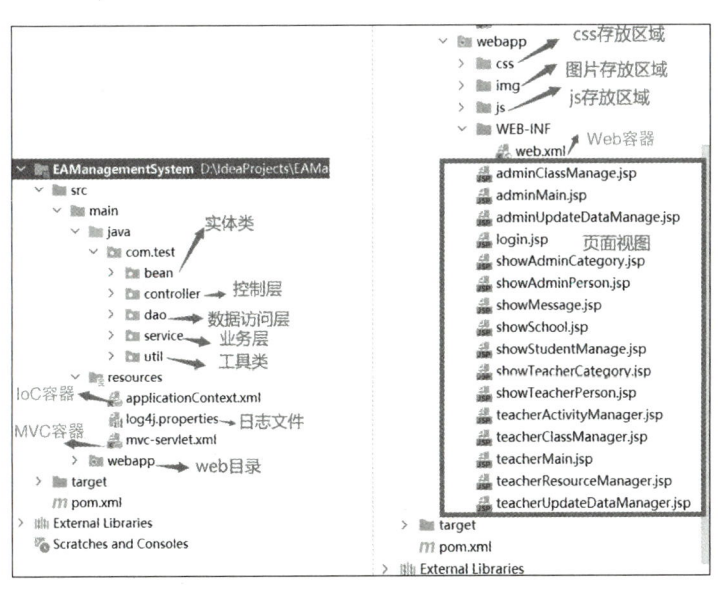

图 15-3　组织结构图

15.4 系统开发及运行环境

教务管理系统开发环境如下。
操作系统：Windows。
Web 服务器：Tomcat 8.0。
Java 开发包：JDK8。
开发工具：IntelliJIDEA 2021.2。
数据库：MySQL 5.5。
浏览器：Google Chrome。

15.5 数据库设计

本系统为教务管理系统，对系统功能进行分析并建立数据库表，如表 15-1~15-6 所示。

表 15-1 管理员表（admin）

字段名	类型	约束	备注
aid	Int	主键自增	编号
aname	Varchar(10)	不为空	管理员姓名
photo	Varchar(50)	默认约束	头像默认 default.jpg
username	Varchar(11)	唯一	登录账号
password	Varchar(6)	不为空，默认约束	默认 666666
remark	Varchar(50)		其他

表 15-2 教师表（teacher）

字段名	类型	约束	备注
tid	Int	主键自增	编号
tname	Varchar(10)	不为空	教师姓名
photo	Varchar(50)	默认约束	头像默认 default.jpg
username	Varchar(11)	唯一	登录账号
password	Varchar(6)	不为空，默认约束	默认 666666
gender	Int	默认 1	性别：1. 男；0. 女
status	Int	默认 1	状态：1. 在职；2. 离职；3. 开除；4. 休假

续表

字段名	类型	约束	备注
islogin	Int	默认1	是否可登录：1. 是；2. 否
phone	Varchar(11)	不为空	手机号码
address	Varchar(50)		家庭住址
remark	Varchar(50)		其他

表15-3 班级表（tbclass）

字段名	类型	约束	备注
cid	Int	主键自增	编号
cname	Varchar(10)	不为空	班级名称
cdesc	Varchar(100)		班级简介
teacher	Int	外键	班主任编号
stucount	Int	默认0	班级人数，不需要手动更新，在添加学生时自动更新
remark	Varchar(50)		其他

表15-4 学生表（student）

字段名	类型	约束	备注
stuid	Varchar(16)	唯一	学生学号，学号格式为s+年份+月份+日期+时分秒+随机数，例如s202203081234566
sname	Varchar(10)	不为空	学生姓名
gender	Int	不为空	性别：1. 男；0. 女
phone	Varchar(11)	不为空	手机号码
address	Varchar(50)		家庭住址
clsid	Int	外键	所在班级id
status	Int	默认1	状态：1. 在籍；2. 毕业；3. 开除；4. 休假
remark	Varchar(50)		其他

表15-5 活动表（tbactivity）

字段名	类型	约束	备注
id	Int	主键自增	活动id
subject	Varchar(20)	不为空	活动主题

续表

字段名	类型	约束	备注
content	Varchar(500)		活动内容
clsid	Int	外键	活动班级 id
starttime	Varchar(10)	不为空	活动时间,格式为 yyyy-mm-dd
remark	Varchar(50)		其他

表 15-6 资料表(tbresource)

字段名	类型	约束	备注
id	Int	主键自增	资料 id
clsid	Int	外键	资料班级 id
filename	Varchar(15)	不为空	资料名称
url	Varchar(100)	不为空	下载地址
remark	Varchar(50)		其他

15.6 系统环境搭建

15.6.1 jar 包依赖

由于本系统使用的是 SSM 框架开发,因此需要依赖这三大框架的 jar 包。除此之外,项目中还涉及数据库连接、JSTL 标签等。整个系统所需要依赖的 jar 如下 pom.xml 文件所示。

```
<?xml version="1.0" encoding="UTF-8"?>
<project xmlns="http://maven.apache.org/POM/4.0.0" xmlns:xsi="http://www.w3.org/2001/XMLSchema-instance" xsi:schemaLocation="http://maven.apache.org/POM/4.0.0 http://maven.apache.org/maven-v4_0_0.xsd">
    <modelVersion>4.0.0</modelVersion>
    <packaging>war</packaging>
    <name>EAManagementSystem</name>
    <groupId>com.test</groupId>
    <artifactId>EAManagementSystem</artifactId>
    <version>1.0-SNAPSHOT</version>
    <properties>
        <project.build.sourceEncoding>UTF-8</project.build.sourceEncoding>
        <spring.version>4.0.2.RELEASE</spring.version>
        <mybatis.version>3.2.8</mybatis.version>
        <slf4j.version>1.7.12</slf4j.version>
        <log4j.version>1.2.17</log4j.version>
        <jackson.version>2.9.3</jackson.version>
    </properties>
```

```xml
<dependencies>
<!-- Spring 框架包 start -->
<dependency>
<groupId>org.springframework</groupId>
<artifactId>spring-test</artifactId>
<version>${spring.version}</version>
</dependency>
<dependency>
<groupId>org.springframework</groupId>
<artifactId>spring-core</artifactId>
<version>${spring.version}</version>
</dependency>
<dependency>
<groupId>org.springframework</groupId>
<artifactId>spring-oxm</artifactId>
<version>${spring.version}</version>
</dependency>
<dependency>
<groupId>org.springframework</groupId>
<artifactId>spring-tx</artifactId>
<version>${spring.version}</version>
</dependency>
<dependency>
<groupId>org.springframework</groupId>
<artifactId>spring-jdbc</artifactId>
<version>${spring.version}</version>
</dependency>
<dependency>
<groupId>org.springframework</groupId>
<artifactId>spring-aop</artifactId>
<version>${spring.version}</version>
</dependency>
<dependency>
<groupId>org.springframework</groupId>
<artifactId>spring-context</artifactId>
<version>${spring.version}</version>
</dependency>
<dependency>
<groupId>org.springframework</groupId>
<artifactId>spring-context-support</artifactId>
<version>${spring.version}</version>
</dependency>
<dependency>
<groupId>org.springframework</groupId>
<artifactId>spring-expression</artifactId>
<version>${spring.version}</version>
</dependency>
<dependency>
```

```xml
<groupId>org.springframework</groupId>
<artifactId>spring-orm</artifactId>
<version>${spring.version}</version>
</dependency>
<dependency>
<groupId>org.springframework</groupId>
<artifactId>spring-web</artifactId>
<version>${spring.version}</version>
</dependency>
<dependency>
<groupId>org.springframework</groupId>
<artifactId>spring-webmvc</artifactId>
<version>${spring.version}</version>
</dependency>
<!-- Spring 框架包 end -->
<!-- MyBatis 框架包 start -->
<dependency>
<groupId>org.mybatis</groupId>
<artifactId>mybatis</artifactId>
<version>${mybatis.version}</version>
</dependency>
<dependency>
<groupId>org.mybatis</groupId>
<artifactId>mybatis-spring</artifactId>
<version>1.2.2</version>
</dependency>
<!-- MyBatis 框架包 end -->
<!-- 分页插件 -->
<!-- https://mvnrepository.com/artifact/com.github.pagehelper/pagehelper -->
<dependency>
<groupId>com.github.pagehelper</groupId>
<artifactId>pagehelper</artifactId>
<version>5.1.2</version>
</dependency>
<!-- 分页插件 -->
<!-- 数据库驱动 -->
<dependency>
<groupId>mysql</groupId>
<artifactId>mysql-connector-java</artifactId>
<version>5.1.35</version>
</dependency>
<!-- 导入 dbcp 的 jar 包，用来在 applicationContext.xml 中配置数据库 -->
<dependency>
<groupId>commons-dbcp</groupId>
<artifactId>commons-dbcp</artifactId>
<version>1.4</version>
</dependency>
<!-- JSTL 标签类 -->
```

```xml
<dependency>
    <groupId>jstl</groupId>
    <artifactId>jstl</artifactId>
    <version>1.2</version>
</dependency>
<!-- log start -->
<dependency>
    <groupId>log4j</groupId>
    <artifactId>log4j</artifactId>
    <version>${log4j.version}</version>
</dependency>
<dependency>
    <groupId>org.slf4j</groupId>
    <artifactId>slf4j-api</artifactId>
    <version>${slf4j.version}</version>
</dependency>
<dependency>
    <groupId>org.slf4j</groupId>
    <artifactId>slf4j-log4j12</artifactId>
    <version>${slf4j.version}</version>
</dependency>
<!-- log END -->
<!-- JSON -->
<!-- 格式化对象,方便输出日志 -->
<dependency>
    <groupId>com.alibaba</groupId>
    <artifactId>fastjson</artifactId>
    <version>1.2.6</version>
</dependency>
<dependency>
    <groupId>org.codehaus.jackson</groupId>
    <artifactId>jackson-mapper-asl</artifactId>
    <version>1.9.13</version>
</dependency>
<!--JSON -->
<dependency>
    <groupId>com.fasterxml.jackson.core</groupId>
    <artifactId>jackson-core</artifactId>
    <version>${jackson.version}</version>
</dependency>
<dependency>
    <groupId>com.fasterxml.jackson.core</groupId>
    <artifactId>jackson-databind</artifactId>
    <version>${jackson.version}</version>
</dependency>
<dependency>
    <groupId>com.fasterxml.jackson.core</groupId>
```

```xml
    <artifactId>jackson-annotations</artifactId>
    <version>${jackson.version}</version>
</dependency>
<!-- JSON -->
<!-- 上传组件包 start -->
<dependency>
    <groupId>commons-fileupload</groupId>
    <artifactId>commons-fileupload</artifactId>
    <version>1.3.1</version>
</dependency>
<dependency>
    <groupId>commons-io</groupId>
    <artifactId>commons-io</artifactId>
    <version>2.4</version>
</dependency>
<dependency>
    <groupId>commons-codec</groupId>
    <artifactId>commons-codec</artifactId>
    <version>1.10</version>
</dependency>
<!-- 解析excel文档 -->
<!-- https://mvnrepository.com/artifact/org.apache.poi/poi -->
<dependency>
    <groupId>org.apache.poi</groupId>
    <artifactId>poi</artifactId>
    <version>3.15</version>
</dependency>
<!-- https://mvnrepository.com/artifact/org.apache.poi/poi-ooxml -->
<dependency>
    <groupId>org.apache.poi</groupId>
    <artifactId>poi-ooxml</artifactId>
    <version>3.15</version>
</dependency>
<!-- 上传组件包 end -->
<!--单元测试的依赖-->
<dependency>
    <groupId>junit</groupId>
    <artifactId>junit</artifactId>
    <version>4.12</version>
    <scope>test</scope>
</dependency>
<dependency>
    <groupId>org.hamcrest</groupId>
    <artifactId>hamcrest-library</artifactId>
    <version>1.3</version>
    <scope>test</scope>
</dependency>
```

```xml
<dependency>
<groupId>org.mockito</groupId>
<artifactId>mockito-core</artifactId>
<version>2.19.0</version>
<scope>test</scope>
</dependency>
<dependency>
<groupId>javax.servlet</groupId>
<artifactId>javax.servlet-api</artifactId>
<version>3.1.0</version>
<scope>provided</scope>
</dependency>
<!-- JavaEE 包 -->
<dependency>
<groupId>javax</groupId>
<artifactId>javaee-api</artifactId>
<version>7.0</version>
<scope>provided</scope>
</dependency>
</dependencies>
<build>
<plugins>
<plugin>
<groupId>org.apache.maven.plugins</groupId>
<artifactId>maven-compiler-plugin</artifactId>
<version>2.3.2</version>
<configuration>
<source>1.8</source>
<target>1.8</target>
<encoding>utf-8</encoding>
</configuration>
</plugin>
<plugin>
<artifactId>maven-war-plugin</artifactId>
<version>2.6</version>
<configuration>
<failOnMissingWebXml>false</failOnMissingWebXml>
</configuration>
</plugin>
<plugin>
<artifactId>maven-surefire-plugin</artifactId>
<version>2.19</version>
<dependencies>
<dependency>
<groupId>org.junit.platform</groupId>
<artifactId>junit-platform-surefire-provider</artifactId>
<version>1.0.0</version>
```

```xml
            </dependency>
          </dependencies>
        </plugin>
      </plugins>
      <resources>
        <resource>
          <directory>src/main/java</directory>
          <includes>
            <include>**/*.xml</include>
          </includes>
          <filtering>true</filtering>
        </resource>
        <resource>
          <directory>src/main/resources</directory>
          <includes>
            <include>**/*.*</include>
          </includes>
        </resource>
      </resources>
    </build>
</project>
```

15.6.2 数据库资源

通过 SQLyog 登录数据库后,创建一个名称为"ea"的数据库,并选择使用该数据库。通过如下指令创建数据库表。

```sql
CREATE DATABASE IF NOT EXISTS 'ea';
USE 'ea';
DROP TABLE IF EXISTS 'admin';
CREATE TABLE 'admin' (
  'aid' int(11) NOT NULL auto_increment,
  'aname' varchar(10) NOT NULL,
  'photo' varchar(50) default 'default.jpg',
  'username' varchar(11) default NULL,
  'PASSWORD' varchar(6) NOT NULL default '666666',
  'remark' varchar(50) default NULL,
  PRIMARY KEY ('aid'),
  UNIQUE KEY 'username' ('username')
) ENGINE=InnoDB DEFAULT CHARSET=utf8;

DROP TABLE IF EXISTS 'student';
CREATE TABLE 'student' (
  'stuid' varchar(16) default NULL,
  'sname' varchar(10) NOT NULL,
```

```sql
'gender' int(11) NOT NULL,
'phone' varchar(11) NOT NULL,
'address' varchar(50) default NULL,
'clsid' int(11) default NULL,
'STATUS' int(11) default '1',
'remark' varchar(50) default NULL,
UNIQUE KEY 'stuid' ('stuid')
) ENGINE=InnoDB DEFAULT CHARSET=utf8;

DROP TABLE IF EXISTS 'tbactivity';
CREATE TABLE 'tbactivity' (
  'id' int(11) NOT NULL auto_increment,
  'SUBJECT' varchar(20) NOT NULL,
  'content' varchar(500) default NULL,
  'clsid' int(11) default NULL,
  'starttime' varchar(10) NOT NULL,
  'remark' varchar(50) default NULL,
  PRIMARY KEY ('id')
) ENGINE=InnoDB DEFAULT CHARSET=utf8;

DROP TABLE IF EXISTS 'tbclass';
CREATE TABLE 'tbclass' (
  'cid' int(11) NOT NULL auto_increment,
  'cname' varchar(10) NOT NULL,
  'cdesc' varchar(100) default NULL,
  'teacher' int(11) default NULL,
  'stucount' int(11) default '0',
  'remark' varchar(50) default NULL,
  PRIMARY KEY ('cid')
) ENGINE=InnoDB DEFAULT CHARSET=utf8;
DROP TABLE IF EXISTS 'tbresource';
CREATE TABLE 'tbresource' (
  'id' int(11) NOT NULL auto_increment,
  'clsid' int(11) default NULL,
  'filename' varchar(15) NOT NULL,
  'url' varchar(255) NOT NULL,
  'remark' varchar(50) default NULL,
  PRIMARY KEY ('id')
) ENGINE=InnoDB DEFAULT CHARSET=utf8;
DROP TABLE IF EXISTS 'teacher';
CREATE TABLE 'teacher' (
  'tid' int(11) NOT NULL auto_increment,
  'tname' varchar(10) NOT NULL,
  'photo' varchar(50) default 'default.jpg',
  'username' varchar(11) default NULL,
  'PASSWORD' varchar(6) NOT NULL default '666666',
  'gender' int(11) default '1',
  'STATUS' int(11) default '1',
```

```
  'islogin' int(11) default '1',
  'phone' varchar(11) NOT NULL,
  'address' varchar(50) default NULL,
  'remark' varchar(50) default NULL,
  PRIMARY KEY  ('tid'),
  UNIQUE KEY 'username' ('username')
) ENGINE=InnoDB DEFAULT CHARSET=utf8;
```

15.6.3 准备项目环境

创建项目(EAManagementSystem),并且在 pom.xml 文件中导入项目所需要 jar 依赖。
在 resources 下编写 IoC 配置文件(applicationContext.xml)。

```xml
<?xml version="1.0" encoding="UTF-8"?>
<beans xmlns="http://www.springframework.org/schema/beans"
xmlns:xsi="http://www.w3.org/2001/XMLSchema-instance"
xmlns:context="http://www.springframework.org/schema/context"
xsi:schemaLocation="http://www.springframework.org/schema/beans
    http://www.springframework.org/schema/beans/spring-beans-4.0.xsd
    http://www.springframework.org/schema/context
http://www.springframework.org/schema/context/spring-context-4.0.xsd">
<context:component-scan base-package="com.test.service"></context:component-scan>
<bean id="dataSource" class="org.springframework.jdbc.datasource.DriverManagerDataSource">
    <property name="driverClassName" value="com.mysql.jdbc.Driver"></property>
    <property name="url" value="jdbc:mysql://localhost:3306/ea"></property>
    <property name="username" value="root"></property>
    <property name="password" value="root"></property>
</bean>
<bean id="ssf" class="org.mybatis.spring.SqlSessionFactoryBean">
    <property name="dataSource" ref="dataSource"></property>
    <property name="mapperLocations" value="classpath:com/test/dao/*-mapper.xml"></property>
    <property name="plugins">
    <array>
    <bean class="com.github.pagehelper.PageInterceptor">
    <property name="properties">
    <value></value>
    </property>
    </bean>
    </array>
    </property>
</bean>
<bean id="msc" class="org.mybatis.spring.mapper.MapperScannerConfigurer">
    <property name="sqlSessionFactoryBeanName" value="ssf"></property>
    <property name="basePackage" value="com.test.dao"></property>
</bean>
</beans>
```

第15章 教务管理系统

在 resources 下编写 Spring MVC 配置文件(mvc-servlet.xml)。

```xml
<beans xmlns="http://www.springframework.org/schema/beans"
    xmlns:xsi="http://www.w3.org/2001/XMLSchema-instance" xmlns:mvc="http://www.springframework.org/schema/mvc"
    xmlns:context="http://www.springframework.org/schema/context"
    xmlns:aop="http://www.springframework.org/schema/aop" xmlns:tx="http://www.springframework.org/schema/tx"
    xsi:schemaLocation="http://www.springframework.org/schema/beans
    http://www.springframework.org/schema/beans/spring-beans-3.2.xsd
    http://www.springframework.org/schema/mvc
    http://www.springframework.org/schema/mvc/spring-mvc-3.2.xsd
    http://www.springframework.org/schema/context
    http://www.springframework.org/schema/context/spring-context-3.2.xsd
    http://www.springframework.org/schema/aop
    http://www.springframework.org/schema/aop/spring-aop-3.2.xsd
    http://www.springframework.org/schema/tx
    http://www.springframework.org/schema/tx/spring-tx-3.2.xsd">
    <context:component-scan base-package="com.test.controller"></context:component-scan>
    <!--
    配置视图解析器
    在返回视图的时候自动加前缀以及后缀
    -->
    <!-- <bean class="org.springframework.web.servlet.view.InternalResourceViewResolver">
    <property name="prefix" value="/"></property>前缀
    <property name="suffix" value=".jsp"></property>后缀
    </bean> -->
    <mvc:annotation-driven/>
    <!-- 上传配置-->
    <bean id="multipartResolver" class="org.springframework.web.multipart.commons.CommonsMultipartResolver">
    <!-- 上传文件最大限制-->
    <property name="maxUploadSize" value="31457280"></property>
    <!-- 上传编码-->
    <property name="defaultEncoding" value="UTF-8"></property>
    </bean>
</beans>
```

在 webapp>WEB-INF 下的 web.xml 中,配置 Spring 的监听器、编码过滤器和 Spring MVC 的前端控制器等信息。

```xml
<?xml version="1.0" encoding="UTF-8"?>
<web-app xmlns:xsi="http://www.w3.org/2001/XMLSchema-instance" xmlns="http://java.sun.com/xml/ns/javaee" xsi:schemaLocation="http://java.sun.com/xml/ns/javaee http://java.sun.com/xml/ns/javaee/web-app_3_0.xsd" id="WebApp_ID" version="3.0">
    <display-name>EAManagementSystem</display-name>
    <welcome-file-list>
    <welcome-file>login.jsp</welcome-file>
```

```xml
</welcome-file-list>
<!-- 配置 Spring MVC 的中央控制器,并加载容器-->
<servlet>
<servlet-name>mvc</servlet-name>
<servlet-class>org.springframework.web.servlet.DispatcherServlet</servlet-class>
<!-- 指定 MVC 容器位置-->
<init-param>
<param-name>contextConfigLocation</param-name>
<param-value>classpath:mvc-servlet.xml</param-value>
</init-param>
</servlet>
<servlet-mapping>
<servlet-name>mvc</servlet-name>
<url-pattern>*.do</url-pattern>
</servlet-mapping>
<listener>
<listener-class>org.springframework.web.context.ContextLoaderListener</listener-class>
</listener>
<context-param>
<param-name>contextConfigLocation</param-name>
<param-value>classpath:applicationContext.xml</param-value>
</context-param>
<filter>
<filter-name>EncodingFilter</filter-name>
<filter-class>org.springframework.web.filter.CharacterEncodingFilter</filter-class>
<init-param>
<param-name>encoding</param-name>
<param-value>UTF-8</param-value>
</init-param>
</filter>
<filter-mapping>
<filter-name>EncodingFilter</filter-name>
<url-pattern>/*</url-pattern>
</filter-mapping>
</web-app>
```

将日志属性配置文件放入 resources 下(log4j.properties)。

```
log4j.rootLogger=DEBUG, stdout
# SqlMap logging configuration...
log4j.logger.com.ibatis=DEBUG
log4j.logger.com.ibatis.common.jdbc.SimpleDataSource=DEBUG
log4j.logger.com.ibatis.sqlmap.engine.cache.CacheModel=DEBUG
log4j.logger.com.ibatis.sqlmap.engine.impl.SqlMapClientImpl=DEBUG
log4j.logger.com.ibatis.sqlmap.engine.builder.xml.SqlMapParser=DEBUG
log4j.logger.com.ibatis.common.util.StopWatch=DEBUG
log4j.logger.java.sql.Connection=DEBUG
```

log4j. logger. java. sql. Statement = DEBUG
log4j. logger. java. sql. PreparedStatement = DEBUG
log4j. logger. java. sql. ResultSet = DEBUG
Console output...
log4j. appender. stdout = org. apache. log4j. ConsoleAppender
log4j. appender. stdout. layout = org. apache. log4j. PatternLayout
log4j. appender. stdout. layout. ConversionPattern = %5p [%t] - %m%n

到此，一个 SSM 的项目环境已经搭建完毕，接下来只需实现功能即可。

15.7 功能实现

15.7.1 创建实体层(Bean)

根据数据表字段，在 com. test. bean 包下创建实体类。

```
package com. test. bean;
public class Admin {
private Integer aid;// 编号
private String aname;// 管理员姓名
private String photo;// 头像默认 default. jpg
private String username;// 登录账号
private String password;// 默认666666
private String remark;// 其他
public Integer getAid( ) {
return aid;
 }
public void setAid( Integer aid) {
this. aid = aid;
 }
public String getAname( ) {
return aname;
 }
public void setAname( String aname) {
this. aname = aname;
 }
public String getPhoto( ) {
return photo;
 }
public void setPhoto( String photo) {
this. photo = photo;
 }
public String getUsername( ) {
```

```java
            return username;
        }
        public void setUsername(String username) {
            this.username = username;
        }
        public String getPassword() {
            return password;
        }
        public void setPassword(String password) {
            this.password = password;
        }
        public String getRemark() {
            return remark;
        }
        public void setRemark(String remark) {
            this.remark = remark;
        }
        @Override
        public String toString() {
            return "Admin [aid=" + aid + ", aname=" + aname + ", photo=" + photo + ", username=" + username + ", password="
                    + password + ", remark=" + remark + "]";
        }
    }

    package com.test.bean;
    public class Student {
        //学号格式为 s+年份+月份+日期+时分秒+随机数,例如 s202203081234566
        private String stuid;// 学生学号
        private String sname;// 学生姓名
        private Integer gender;// 性别:1.男,0.女
        private String phone;// 手机号码
        private String address;// 家庭住址
        private Integer clsid;// 外键
        private Integer status;// 状态:1.在籍,2.毕业,3.开除,4.休假
        private String remark;// 其他
        private TbClass tbclass;
        public String getStuid() {
            return stuid;
        }
        public void setStuid(String stuid) {
            this.stuid = stuid;
        }
        public String getSname() {
            return sname;
        }
        public void setSname(String sname) {
            this.sname = sname;
```

```java
    }
    public Integer getGender() {
        return gender;
    }
    public void setGender(Integer gender) {
        this.gender = gender;
    }
    public String getPhone() {
        return phone;
    }
    public void setPhone(String phone) {
        this.phone = phone;
    }
    public String getAddress() {
        return address;
    }
    public void setAddress(String address) {
        this.address = address;
    }
    public Integer getClsid() {
        return clsid;
    }
    public void setClsid(Integer clsid) {
        this.clsid = clsid;
    }
    public Integer getStatus() {
        return status;
    }
    public void setStatus(Integer status) {
        this.status = status;
    }
    public String getRemark() {
        return remark;
    }
    public void setRemark(String remark) {
        this.remark = remark;
    }
    public TbClass getTbclass() {
        return tbclass;
    }
    public void setTbclass(TbClass tbclass) {
        this.tbclass = tbclass;
    }
    @Override
    public String toString() {
        return "Student [stuid=" + stuid + ", sname=" + sname + ", gender=" + gender + ", phone=" + phone + ", address="
```

```java
        + address + ",clsid=" + clsid+ ",status=" + status + ",remark=" + remark + "]";
    }
}
package com.test.bean;
public class TbActivity{
    private Integer id;// 活动id
    private String subject;// 活动主题
    private String content;// 活动内容
    private Integer clsid;// <外键>活动班级id
    private String starttime;// 活动时间格式为yyyy-mm-dd
    private String remark;// 其他
    private TbClass tbclass;
    public Integer getId(){
        return id;
    }
    public void setId(Integer id){
        this.id = id;
    }
    public String getSubject(){
        return subject;
    }
    public void setSubject(String subject){
        this.subject = subject;
    }
    public String getContent(){
        return content;
    }
    public void setContent(String content){
        this.content = content;
    }
    public Integer getClsid(){
        return clsid;
    }
    public void setClsid(Integer clsid){
        this.clsid = clsid;
    }
    public String getStarttime(){
        return starttime;
    }
    public void setStarttime(String starttime){
        this.starttime = starttime;
    }
    public String getRemark(){
        return remark;
    }
    public void setRemark(String remark){
        this.remark = remark;
```

```java
        }
    public TbClass getTbclass() {
        return tbclass;
    }
    public void setTbclass(TbClass tbclass) {
        this.tbclass = tbclass;
    }
    public TbActivity() {
        super();
        // TODO Auto-generated constructor stub
    }
    public TbActivity(Integer id, String subject, String content, Integer clsid, String starttime, String remark) {
        super();
        this.id = id;
        this.subject = subject;
        this.content = content;
        this.clsid = clsid;
        this.starttime = starttime;
        this.remark = remark;
    }
    @Override
    public String toString() {
        return "TbActivity [id=" + id + ", subject=" + subject + ", content=" + content + ", clsid=" + clsid
                + ", starttime=" + starttime + ", remark=" + remark + "]";
    }
}

package com.test.bean;
import java.util.List;
public class TbClass {
    private Integer cid;// 编号
    private String cname;// 班级姓名
    private String cdesc;// 班级简介
    private Integer teacher;// <外键>班主任编号
    private Integer stucount;// 班级人数不需要手动更新,在添加学生时自动更新
    private String remark;// 其他
    private Teacher tr;//班主任
    private List<Student> stu;
    public Integer getCid() {
        return cid;
    }
    public void setCid(Integer cid) {
        this.cid = cid;
    }
    public String getCname() {
        return cname;
    }
    public void setCname(String cname) {
```

```java
        this.cname = cname;
    }
    public String getCdesc() {
        return cdesc;
    }
    public void setCdesc(String cdesc) {
        this.cdesc = cdesc;
    }
    public Integer getTeacher() {
        return teacher;
    }
    public void setTeacher(Integer teacher) {
        this.teacher = teacher;
    }
    public Integer getStucount() {
        return stucount;
    }
    public void setStucount(Integer stucount) {
        this.stucount = stucount;
    }
    public String getRemark() {
        return remark;
    }
    public void setRemark(String remark) {
        this.remark = remark;
    }
    public Teacher getTr() {
        return tr;
    }
    public void setTr(Teacher tr) {
        this.tr = tr;
    }
    public List<Student> getStu() {
        return stu;
    }
    public void setStu(List<Student> stu) {
        this.stu = stu;
    }
    @Override
    public String toString() {
        return "TbClass [cid=" + cid + ", cname=" + cname + ", cdesc=" + cdesc + ", teacher=" + teacher + ", stucount="
                + stucount + ", remark=" + remark + ", stu=" + stu + "]";
    }
}
package com.test.bean;
public class TbResource{
```

```java
private Integer id;// 资料 id
private Integer clsid;// <外键>资料班级 id
private String filename;// 资料名称
private String url;// 下载地址
private String remark;// 其他
private TbClass tbclass;
public Integer getId() {
return id;
}
public void setId(Integer id) {
this.id = id;
}
public Integer getClsid() {
return clsid;
}
public void setClsid(Integer clsid) {
this.clsid = clsid;
}
public String getFilename() {
return filename;
}
public void setFilename(String filename) {
this.filename = filename;
}
public String getUrl() {
return url;
}
public void setUrl(String url) {
this.url = url;
}
public String getRemark() {
return remark;
}
public void setRemark(String remark) {
this.remark = remark;
}
public TbClass getTbclass() {
return tbclass;
}
public void setTbclass(TbClass tbclass) {
this.tbclass = tbclass;
}
@Override
public String toString() {
return "TbResource [id=" + id + ", clsid=" + clsid+ ", filename=" + filename + ", url=" + url+ ", remark="
+ remark + "]";
}
```

```java
}
package com.test.bean;
public class Teacher {
    private Integer tid;// 编号
    private String tname;// 管理员姓名
    private String photo;// 头像默认default.jpg
    private String username;// 登录账号
    private String password;// 默认666666
    private Integer gender;// 性别:1.男,0.女
    private Integer status;// 状态:1.在职,2.离职,3.开除,4.休假
    private Integer islogin;// 是否可登录:1.是,2.否
    private String phone;// 联系手机号码
    private String address;// 家庭住址
    private String remark;// 其他
    public Integer getTid() {
        return tid;
    }
    public void setTid(Integer tid) {
        this.tid = tid;
    }
    public String getTname() {
        return tname;
    }
    public void setTname(String tname) {
        this.tname = tname;
    }
    public String getPhoto() {
        return photo;
    }
    public void setPhoto(String photo) {
        this.photo = photo;
    }
    public String getUsername() {
        return username;
    }
    public void setUsername(String username) {
        this.username = username;
    }
    public String getPassword() {
        return password;
    }
    public void setPassword(String password) {
        this.password = password;
    }
    public Integer getGender() {
        return gender;
    }
```

```java
public void setGender(Integer gender){
    this.gender = gender;
}
public Integer getStatus(){
    return status;
}
public void setStatus(Integer status){
    this.status = status;
}
public Integer getIslogin(){
    return islogin;
}
public void setIslogin(Integer islogin){
    this.islogin = islogin;
}
public String getPhone(){
    return phone;
}
public void setPhone(String phone){
    this.phone = phone;
}
public String getAddress(){
    return address;
}
public void setAddress(String address){
    this.address = address;
}
public String getRemark(){
    return remark;
}
public void setRemark(String remark){
    this.remark = remark;
}
public Teacher(){
    super();
    // TODO Auto-generated constructor stub
}
public Teacher(Integer tid, String tname, String photo, String username, String password, Integer gender, Integer status, Integer islogin, String phone, String address, String remark){
    super();
    this.tid = tid;
    this.tname = tname;
    this.photo = photo;
    this.username = username;
    this.password = password;
    this.gender = gender;
    this.status = status;
```

```
this. islogin = islogin;
this. phone = phone;
this. address = address;
this. remark = remark;
    }
public Teacher(String tname, String photo, String username, String password, Integer gender, Integer status,
Integer islogin, String phone, String address, String remark) {
super( );
this. tname = tname;
this. photo = photo;
this. username = username;
this. password = password;
this. gender = gender;
this. status = status;
this. islogin = islogin;
this. phone = phone;
this. address = address;
this. remark = remark;
    }
@ Override
public String toString( ) {
return "Teacher [tid=" + tid+ ", tname=" + tname+ ", photo=" + photo + ", username=" + username
+ ", password=" + password + ", gender=" + gender + ", status=" + status + ", islogin=" + islogin
+ ", phone=" + phone + ", address=" + address + ", remark=" + remark + "]";
    }
}
```

实体层目录结构如图 15-4 所示。

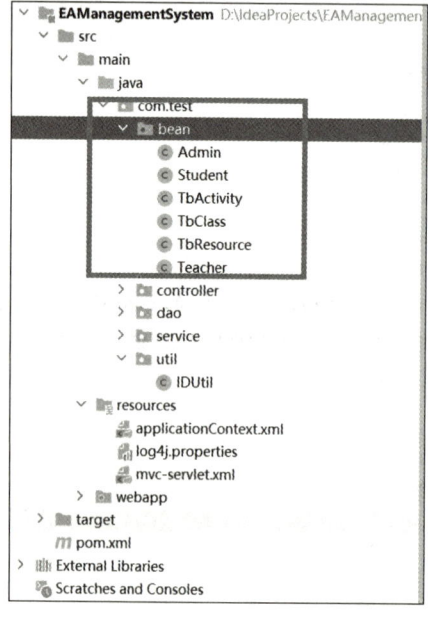

图 15-4　实体层目录结构图

15.7.2 创建工具类(util)

在 com.test.uitl 下创建生成 id 的工具类。

```java
package com.test.util;
import java.text.DateFormat;
import java.text.SimpleDateFormat;
import java.util.Date;
public class IDUtil{
/**
    学生学号格式为 s+年份+月份+日期+时分秒+随机数,例如 s202203081234566
    * @return
    */
public static final String getIDUtil() {
String id = "s";
int num = (int)(Math.random()*10);
DateFormat dateFormat= new SimpleDateFormat("yyyyMMddHHmmss");
Date date= new Date();
        id += dateFormat.format(date) +num;
return id;
    }
/**
    * 获取当前时间
* @return
    */
public static final String getTimeUtil() {
DateFormat dateFormat= new SimpleDateFormat("yyyy-MM-dd");
Date date= new Date();
return dateFormat.format(date);
    }
}
```

工具类目录结构如图 15-5 所示。

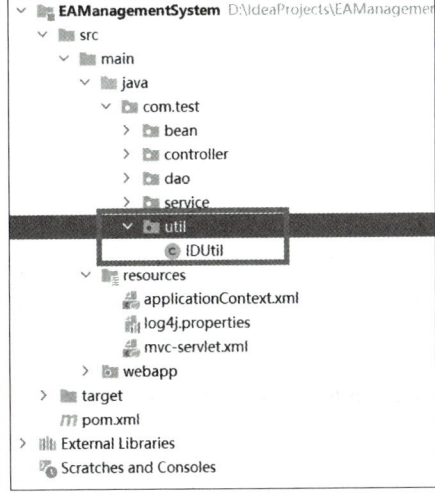

图 15-5 工具类目录结构图

15.7.3 创建数据访问层(DAO)

根据功能需求,在 com.test.dao 包下面创建数据访问接口以及与之对应的映射文件。

管理员数据访问层(IAdminDao)以及对应映射文件(IAdminDao-mapper.xml)如下所示。

```java
package com.test.dao;
import com.test.bean.Admin;
/**
 * 数据访问层接口,管理员完成功能
 */
public interface IAdminDao{
/**
    * 管理员登录
    * @param admin
    */
public Admin adminLogin(Admin admin) throws Exception;
/**
    * 管理员密码重置(666666)
    * @param aid
    */
public Integer adminResertPS(Integer aid) throws Exception;
/**
    * 管理员修改本人资料
    * @param admin
    * @return
    * @throws Exception
    */
public Integer adminUpdateByThis(Admin admin) throws Exception;
/**
    * 获得个人信息
    * @param aid
    */
public Admin queryAdminById(Integer aid) throws Exception;
}
```

```xml
<!DOCTYPE mapper PUBLIC
"-//mybatis.org//DTD Mapper 3.0//EN"
    "http://mybatis.org/dtd/mybatis-3-mapper.dtd">
<mapper namespace="com.test.dao.IAdminDao">
<sqlid="adminMsgAll">aid,aname,photo,username,PASSWORD,remark</sql>
<select id="adminLogin" parameterType="com.test.bean.Admin" resultType="com.test.bean.Admin">
    select * from admin where username=#{username} and password=#{password}
</select>
<update id="adminResertPS" parameterType="int">
        update admin set password=666666 where aid=#{aid}
</update>
```

```xml
<update id="adminUpdateByThis" parameterType="com.test.bean.Admin">
    update admin set aname=#{aname},photo=#{photo},username=#{username},
    password=#{password},remark=#{remark} where aid=#{aid}
</update>
<select id="queryAdminById" resultType="com.test.bean.Admin">
    select * from admin where aid=#{aid}
</select>
</mapper>
```

学生数据访问层(IStudentDao)以及对应映射文件(IStudentDao-mapper.xml)如下所示。

```java
package com.test.dao;
import java.util.List;
import java.util.Map;
import org.apache.ibatis.annotations.Param;
import com.test.bean.Student;
/**
 * 数据访问层接口
 */
public interface IStudentDao{
    /**
     *  分页展示学生信息
     */
    public List<Student>queryAllStudentMessage() throws Exception;
    /**
     *  根据班级学号姓名模糊查询学生
     *
     *  @param clsid
     *         班级id
     *  @param stuid
     *         学号
     * @param sname
     *         姓名
     */
    public List<Student>dimQureyStudentByClassAndSidAndName(@Param("clsid") String clsid,
    @Param("stuid") Integer stuid,@Param("sname") String sname) throws Exception;
//------------------------------------------------------------
    //班主任操作
    /**
     *  添加学生
     *
     *  @param student
     */
    public Integer insertStudent(Student student) throws Exception;
    /**
     *  修改学生信息
     *
```

```java
     * @param student
     */
    public Integer updateStudentMessage(Student student) throws Exception;
    /**
     * 开除学生(不删除仅修改状态)/恢复开除的学生
     *
     * @param status
     * @param stuid
     */
    public Integer expelStudent(@Param("status") Integer status, @Param("stuid") String stuid) throws Exception;
    // ----------------------------------------------
    // 管理员6.学生管理(分页展示学生信息,根据班级学号姓名模糊查询学生)
    // 联合查询
    /**
     * 联合查询分页展示学生信息
     *
     */
    public List<Student> queryAllClassAndStudent(Map<String,Object> map) throws Exception;
    /**
     * 查询学生通过id
     * @param stuid
     *
     */
    public Student queryStudentById(String stuid) throws Exception;
    /**
     * 两表联合查询学生信息通过班级id
     * @param clsid
     */
    public List<Student> queryByStudentByClsid(Integer clsid) throws Exception;
}
```

```xml
<!DOCTYPE mapper PUBLIC
"-//mybatis.org//DTD Mapper 3.0//EN"
    "http://mybatis.org/dtd/mybatis-3-mapper.dtd">
<mapper namespace="com.test.dao.IStudentDao">
<select id="queryAllStudentMessage" resultType="com.test.bean.Student">
        select * from student
</select>
<select id="dimQureyStudentByClassAndSidAndName" parameterType="map" resultType="com.test.bean.Student">
        select * from student
<where>
<if test="clsid!=null">
clsid like <![CDATA["%"]]>${clsid}<![CDATA["%"]]>
</if>
<if test="stuid!=null">
<trim prefix="and">
stuid like <![CDATA["%"]]>${stuid}<![CDATA["%"]]>
</trim>
```

```xml
        </if>
        <if test="sname!=null">
            <trim prefix="and">
                sname like <![CDATA["%"]]>${sname}<![CDATA["%"]]>
            </trim>
        </if>
    </where>
</select>
<insert id="insertStudent" parameterType="com.test.bean.Student">
    insert into student(stuid,sname,gender,phone,address,clsid,status,remark)
    values(#{stuid},#{sname},#{gender},#{phone},#{address},#{clsid},#{status},#{remark})
</insert>
<update id="updateStudentMessage" parameterType="com.test.bean.Student">
    update student set sname=#{sname},gender=#{gender},phone=#{phone},   address=#{address},remark=#{remark},clsid=#{clsid},status=#{status}
    where stuid=#{stuid}
</update>
<update id="expelStudent" parameterType="map">
    update student set status=#{status} where stuid=#{stuid}
</update>
<resultMap type="com.test.bean.Student" id="stu">
    <result column="stuid" property="stuid"/>
    <result column="sname" property="sname"/>
    <result column="gender" property="gender"/>
    <result column="phone" property="phone"/>
    <result column="clsid" property="clsid"/>
    <result column="status" property="status"/>
    <result column="remark" property="remark"/>
    <!--
    一对一使用 association 标签
            property:表示需要查询出来的属性
            select:指向一个查询方法用于查询级联属性
      column:使用指向一个级联字段,表示将该字段当成参数传给 select 指向的方法,
    查询出一个对象,赋值给 property
            -->
    <association property="tbclass"
    select="com.test.dao.ITbClassDao.queryById"
    column="clsid">
    </association>
</resultMap>
<select id="queryAllClassAndStudent" parameterType="map" resultMap="stu">
    select * from student
    <where>
        <if test="clsid!=null">
            clsid like <![CDATA["%"]]>${clsid}<![CDATA["%"]]>
        </if>
        <if test="stuid!=null">
            <trim prefix="and">
```

```
            stuid like <![CDATA["%"]]>${stuid}<![CDATA[%"]]>
        </trim>
    </if>
    <if test="sname!=null">
        <trim prefix=" and ">
            sname like <![CDATA["%"]]>${sname}<![CDATA[%"]]>
        </trim>
    </if>
    </where>
</select>
<select id="queryStudentById" resultType="com.test.bean.Student">
        select * from student where stuid=#{stuid}
</select>
<resultMap type="com.test.bean.Student" id="byStuByClsid">
<!--
一对一使用 association 标签
            property:表示需要查询出来的属性
            select:指向一个查询方法用于查询级联属性
            column:使用指向一个级联字段,表示将该字段当成参数传给 select 指向的方法,查询出一个对象,赋值给 property
            -->
<association property="tbclass"
select="com.test.dao.ITbClassDao.queryById"
column="clsid">
</association>
</resultMap>
<select id="queryByStudentByClsid" resultMap="byStuByClsid">
        select * from student where clsid=#{clsid}
</select>
</mapper>
```

活动数据访问层(ITbActivityDao)以及对应映射文件(ITbActivityDao-mapper.xml)如下所示。

```
package com.test.dao;
import java.util.List;
import com.test.bean.TbActivity;
/**
 * 数据访问层接口活动管理
 */
public interface ITbActivityDao{
//----------------------------------------------------------
    //班主任操作
    /**
     * 创建活动
     * @param activity
     */
    public Integer createActivity(TbActivity activity) throws Exception;
    /**
```

```
 * 修改活动
 * @param activity
 */
public Integer updateActivity(TbActivity activity) throws Exception;
/**
 * 分页展示活动
 */
public List<TbActivity> queryAllActivity() throws Exception;
/**
 * 单击活动显示活动详情
 * @param id
 */
public TbActivity queryActivityMessage(Integer id) throws Exception;
}
```

```xml
<!DOCTYPE mapper PUBLIC
"-//mybatis.org//DTD Mapper 3.0//EN"
    "http://mybatis.org/dtd/mybatis-3-mapper.dtd">
<mapper namespace="com.test.dao.ITbActivityDao">
<insert id="createActivity" parameterType="com.test.bean.TbActivity">
        insert into tbactivity(subject,content,clsid,starttime,remark)
        values(#{subject},#{content},#{clsid},#{starttime},#{remark})
</insert>
<update id="updateActivity" parameterType="com.test.bean.TbActivity">
        update tbactivity set subject=#{subject},content=#{content},clsid=#{clsid},
        remark=#{remark} where id=#{id}
</update>
<resultMap type="com.test.bean.TbActivity" id="tbAct">
<!--
一对一使用 association 标签
            property:表示需要查询出来的属性
            select:指向一个查询方法用于查询级联属性
            column:使用指向一个级联字段,表示将该字段当成参数传给 select 指向的方法,查询出一个对象,赋值给 property
                -->
<association property="tbclass"
select="com.test.dao.ITbClassDao.queryById"
column="clsid">
</association>
</resultMap>
<select id="queryAllActivity" resultMap="tbAct">
        select * from tbActivity
</select>
<resultMap type="com.test.bean.TbActivity" id="dataTbAct">
<association property="tbclass"
select="com.kd.dao.ITbClassDao.queryById"
column="clsid">
</association>
</resultMap>
```

```xml
<select id="queryActivityMessage" resultMap="dataTbAct">
    select * from tbActivity where id=#{id}
</select>
</mapper>
```

班级数据访问层(ITbClassDao)以及对应映射文件(ITbClassDao-mapper.xml)如下所示。

```java
package com.test.dao;
import java.util.List;
import org.apache.ibatis.annotations.Param;
import com.test.bean.TbClass;
/**
 * 数据访问层班级管理
 * @author Administrator
 */
public interface ITbClassDao{
//------------------------------------------------------------
    //管理员操作
    /**
     * 分页展示班级信息
     */
    public List<TbClass> dimQueryTbClassMessage() throws Exception;
    /**
     * 开设班级
     * @param tbclass
     */
    public Integer insertTbClass(TbClass tbclass) throws Exception;
    /**
     * 修改班级班主任
     * @param teacher
     * @param cid
     */
    public Integer updateTbClass(@Param("teacher")Integer teacher,@Param("cid")Integer cid) throws Exception;
    /**
     * 通过班级名称查询班级
     */
    public TbClass queryByClassName() throws Exception;
//------------------------------------------------------------
    //班主任操作
    /**
     * 分页展示班级信息
     */
    public List<TbClass> queryAllTbClass() throws Exception;
    /**
     * 单击班级在下方分页显示班级学生信息
     * @param cid
     */
    public List<TbClass> queryTbClassStudentMessage(Integer cid) throws Exception;
```

```
//---------------
    /**
    * 通过学生关联id查询班级
    * @param cid
    */
    public TbClass queryById(Integer cid) throws Exception;
    /**
    * 查询班级人数
    * @param cid
    */
    public Integer queryStuCountById(Integer cid) throws Exception;
    /**
    * 修改班级人数
    * @param cid
    * @param count
    */
    public Integer updateClassPersonNumber(@Param("cid") Integer cid,@Param("count") Integer count)  throws Exception;
    /**
    * 修改班级资料
    * @param tbclass
    */
    public Integer updateClassMessage(TbClass tbclass) throws Exception;
    /**
    * 修改班级信息,通过班级id查询
    * @param cid
    */
    public TbClass queryClassById(Integer cid) throws Exception;
}
<!DOCTYPE mapper PUBLIC
"-//mybatis.org//DTD Mapper 3.0//EN"
    "http://mybatis.org/dtd/mybatis-3-mapper.dtd">
<mapper namespace="com.test.dao.ITbClassDao">
<resultMap type="com.test.bean.TbClass" id="tc">
<result column="cid" property="cid"/>
<result column="cname" property="cname"/>
<result column="cdesc" property="cdesc"/>
<result column="teacher" property="teacher"/>
<result column="stucount" property="stucount"/>
<result column="remark" property="remark"/>
<association property="tr"
select="com.test.dao.ITeacherDao.queryById"
column="teacher">
</association>
</resultMap>
<select id="dimQueryTbClassMessage" resultMap="tc">
        select * from tbclass
</select>
```

```xml
<insert id="insertTbClass" parameterType="com.test.bean.TbClass">
        insert into tbclass(cname,cdesc,teacher,stucount,remark)
    values(#{cname},#{cdesc},#{teacher},#{stucount},#{remark});
</insert>
<update id="updateTbClass">
        update tbclass set teacher=#{teacher} where cid=#{cid}
</update>
<select id="queryByClassName" parameterType="string">
        select * from tbclass where cname=#{cname}
</select>
<select id="queryAllTbClass" resultType="com.test.bean.TbClass">
        select * from tbclass
</select>
<select id="queryTbClassStudentMessage" parameterType="int">
        select s.* from tbclasst,student s where t.cid=s.clsidadncid=#{cid};
</select>
<resultMap type="com.test.bean.TbClass" id="tbData">
<association property="tr"
select="com.test.dao.ITeacherDao.queryById"
column="teacher">
</association>
</resultMap>
<select id="queryById" resultMap="tbData">
        select * from tbclass where cid=#{cid}
</select>
<select id="queryStuCountById" resultType="int">
        select stucount from tbclass where cid=#{cid}
</select>
<update id="updateClassPersonNumber">
        update tbclass set stucount=#{count} where cid=#{cid}
</update>
<update id="updateClassMessage" parameterType="com.test.bean.TbClass">
        update tbclass set cname=#{cname},cdesc=#{cdesc},
        teacher=#{teacher},stucount=#{stucount},remark=#{remark}
        where cid=#{cid}
</update>
<select id="queryClassById" resultType="com.test.bean.TbClass">
        select * from tbclass where cid=#{cid}
</select>
</mapper>
```

资料数据访问层(ITbResourceDao)以及对应映射文件(ITbResourceDao-mapper.xml)如下所示。

```java
package com.test.dao;
import java.util.List;
import com.test.bean.TbResource;
/**
 * 数据访问层接口资料表
```

```java
*/
public interface ITbResourceDao{
//----------------------------------------------------------
    //班主任操作
/**
    * 上传班级资料
* @param reource
*/
public Integer uploadTbClassData(TbResource reource) throws Exception;
/**
    * 资料展示
*/
public List<TbResource> queryResource() throws Exception;
/**
    * 查询班级资料通过班级id
    * @param clsid
*/
public List<TbResource> queryByResourceByClsid(Integer clsid) throws Exception;;
/**
    * 修改班级资料
* @param resource
*/
public Integer updateClassData(TbResource resource) throws Exception;
/**
    * 查询资料通过
* @param id
*/
public TbResource queryResourceById(Integer id) throws Exception;
}
```

```xml
<!DOCTYPE mapper PUBLIC
"-//mybatis.org//DTD Mapper 3.0//EN"
    "http://mybatis.org/dtd/mybatis-3-mapper.dtd">
<mapper namespace="com.test.dao.ITbResourceDao">
<insert id="uploadTbClassData" parameterType="com.test.bean.TbResource">
        insert into tbresource(clsid,filename,url,remark)
        values(#{clsid},#{filename},#{url},#{remark})
</insert>
<resultMap type="com.test.bean.TbResource" id="tbr">
<association property="tbclass"
select="com.test.dao.ITbClassDao.queryById"
column="clsid">
</association>
</resultMap>
<select id="queryResource" resultMap="tbr">
        select * from tbresource
</select>
<select id="queryByResourceByClsid" resultType="com.test.bean.TbResource">
        select * from tbresource where clsid=#{clsid}
```

```xml
</select>
<update id="updateClassData" parameterType="com.test.bean.TbResource">
    update tbresource set filename=#{filename},url=#{url},remark=#{remark}
    where id=#{id}
</update>
<select id="queryResourceById" resultType="com.test.bean.TbResource">
    select * from tbresource where id=#{id}
</select>
</mapper>
```

教师数据访问层(ITeacherDao)以及对应映射文件(ITeacherDao-mapper.xml)如下所示。

```java
package com.test.dao;
import java.util.List;
import org.apache.ibatis.annotations.Param;
import com.test.bean.Teacher;
/**
 * 数据访问层接口教师管理*
 */
public interface ITeacherDao{
//-----------------------------------------------------------
    //管理员操作
    /**
     * 分页展示老师信息
     */
    public List<Teacher> queryAllTeacherMessage() throws Exception;
    /**
     * 添加老师信息
     * @param teacher
     */
    public Integer insertTeacher(Teacher teacher) throws Exception;
    /**
     * 开除老师(不删除仅修改状态)/恢复开除老师状态
     * @param tid
     * @param status
     */
    public Integer expelTeacher(@Param("tid")Integer tid,@Param("status")Integer status) throws Exception;
    /**
     * 禁止登录/解除禁止登录
     * @param tid
     * @param islogin
     */
    public Integer forbidTeacherLogin(@Param("tid")Integer tid,@Param("islogin")Integer islogin) throws Exception;
    /**
     * 通过班主任名称查询是否存在
     */
```

```java
public Teacher queryByTeacherName() throws Exception;
//----------------------------------------------------------
    //班主任操作
/**
    * 班主任登录
* @param teacher
*/
public Teacher teacherLogin(Teacher teacher) throws Exception;
/**
    * 班主任修改密码
* @param tid 班主任编号
* @param password 密码
*/
public Integer teacherUpdatePassword(@Param("id") Integer tid,@Param("password") Integer password) throws Exception;
/**
    * 班主任修改个人资料
* @param teacher
*/
public Integer teacherUpdatePersonalData(Teacher teacher) throws Exception;
//----------------------
    /**
    * 管理员操作
* 分页展示班级信息两表联合
* @param tid
*/
public Teacher queryById(Integer tid) throws Exception;
}
```

```xml
<!DOCTYPE mapper PUBLIC
"-//mybatis.org//DTD Mapper 3.0//EN"
    "http://mybatis.org/dtd/mybatis-3-mapper.dtd">
<mapper namespace="com.test.dao.ITeacherDao">
<select id="queryAllTeacherMessage" resultType="com.test.bean.Teacher">
        select * from teacher
</select>
<insert id="insertTeacher" parameterType="com.test.bean.Teacher">
        insert into teacher(tname,photo,username,gender,phone,address,remark)        values(#{tname},#{photo},#{username},#{gender},#{phone},#{address},#{remark})
</insert>
<update id="expelTeacher" parameterType="map">
        update teacher set status=#{status} where tid=#{tid}
</update>
<update id="forbidTeacherLogin" parameterType="map">
        update teacher set islogin=#{islogin} where tid=#{tid}
</update>
<select id="queryByTeacherName" parameterType="string">
        select * from teacher where tname=#{tname}
```

```xml
</select>
<select id="teacherLogin" resultType="com.test.bean.Teacher" parameterType="com.test.bean.Teacher">
    select * from teacher where username=#{username} and password=#{password}
</select>
<update id="teacherUpdatePassword" parameterType="map">
    update teacher set password=#{password} where tid=#{tid}
</update>
<update id="teacherUpdatePersonalData" parameterType="com.test.bean.Teacher">
    update teacher set tname=#{tname},photo=#{photo},username=#{username},gender=#{gender},status=#{status},islogin=#{islogin},phone=#{phone},address=#{address},remark=#{remark} where tid=#{tid}
</update>
<select id="queryById" resultType="com.test.bean.Teacher">
    select * from teacher where tid=#{tid}
</select>
</mapper>
```

数据访问层目录结构如图 15-6 所示。

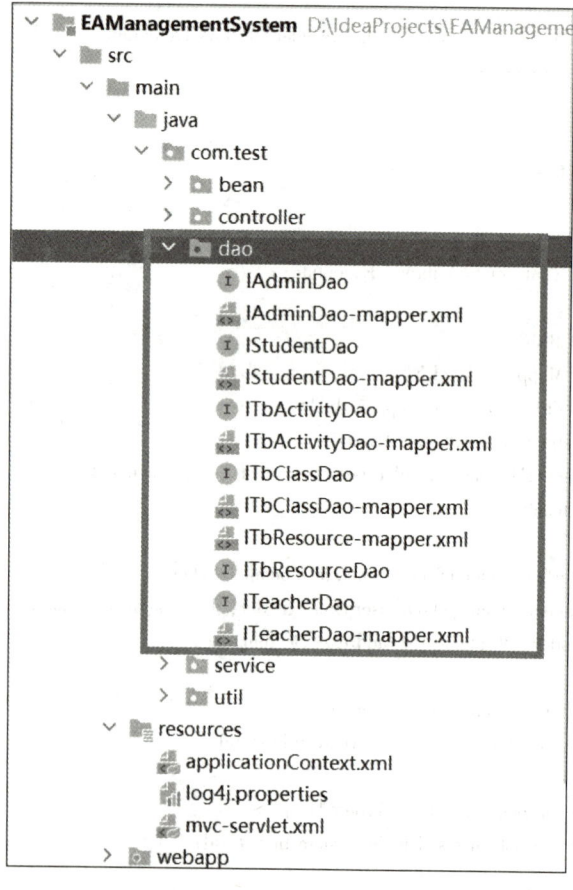

图 15-6　数据访问层目录结构图

15.7.4 创建业务逻辑层(service)

根据功能需求,在 com. test. service 包下面创建服务层接口与之对应的实现类。

管理员服务层接口(IAdminService)以及对应实现类(AdminServiceImpl)如下所示。

```
package com. test. service;
import com. test. bean. Admin;
/**
* 服务层功能
*/
public interface IAdminService{
/**
    * 管理员登录
* @ param admin
*/
public Admin adminLogin( Admin admin) throws Exception;
/**
    * 管理员密码重置(666666)
* @ param aid
*/
public Boolean adminResertPS( Integer aid) throws Exception;
/**
    * 管理员修改本人资料
* @ param admin
*/
public Boolean adminUpdateByThis( Admin admin) throws Exception;
/**
    * 获得个人信息
* @ param aid
*/
public Admin queryAdminById( Integer aid) throws Exception;
}
package com. test. service. impl;
import org. springframework. beans. factory. annotation. Autowired;
import org. springframework. stereotype. Service;
import com. test. bean. Admin;
import com. test. dao. IAdminDao;
import com. test. service. IAdminService;
@ Service
public class AdminServiceImplimplements IAdminService{
@ Autowired
IAdminDaodao;
@ Override
public Admin adminLogin( Admin admin) throws Exception{
return dao. adminLogin( admin);
}
```

```java
@Override
public Boolean adminResertPS(Integer aid) throws Exception{
    return dao.adminResertPS(aid)>0;
}
@Override
public Boolean adminUpdateByThis(Admin admin) throws Exception{
    return dao.adminUpdateByThis(admin)>0;
}
@Override
public Admin queryAdminById(Integer aid) throws Exception{
    return dao.queryAdminById(aid);
}
}
```

学生服务层接口(IStudentService)以及对应实现类(StudentServiceImpl)如下所示。

```java
package com.test.service;
import java.util.List;
import java.util.Map;
import com.github.pagehelper.PageInfo;
import com.test.bean.Student;
/**
 * 服务层学生管理
 */
public interface IStudentService{
    /**
     * 分页展示学生信息
     */
    public PageInfo<Student> queryAllStudentMessage(Integer pageNum,IntegerpageSize) throws Exception;
    /**
     * 根据班级学号姓名模糊查询学生
     * @param clsid 班级id
     * @param stuid 学号
     * @param sname 姓名
     */
    public List<Student> dimQureyStudentByClassAndSidAndName(
        String clsid,Integerstuid,Stringsname
    ) throws Exception;
    //---------------------------------------------------------
    //班主任操作
    /**
     * 添加学生
     * @param student
     */
    public Boolean insertStudent(Student student) throws Exception;
    /**
     * 修改学生信息
```

```java
 * @param student
 */
public Boolean updateStudentMessage(Student student) throws Exception;
/**
 * 开除学生(不删除仅修改状态)/恢复开除的学生
 * @param status
 * @param stuid
 */
public Boolean expelStudent(Integer status,String stuid) throws Exception;
/**
 * 联合查询分页展示学生信息
 */
public PageInfo<Student> queryAllClassAndStudent(Integer pageNum,Integer pageSize,
Map<String,Object> map) throws Exception;
/**
 * 通过id查询学生
 * @param stuid
 */
public Student queryStudentById(String stuid) throws Exception;
/**
 * 通过班级id两表联合查询学生信息
 * @param clsid
 */
public PageInfo<Student> queryByStudentByClsid(Integer clsid,Integer pageNum,Integer pageSize) throws Exception;
}
package com.test.service.impl;
import java.util.List;
import java.util.Map;
import org.springframework.beans.factory.annotation.Autowired;
import org.springframework.stereotype.Service;
import com.github.pagehelper.PageHelper;
import com.github.pagehelper.PageInfo;
import com.test.bean.Student;
import com.test.dao.IStudentDao;
import com.test.dao.ITbClassDao;
import com.test.service.IStudentService;
import com.test.util.IDUtil;
@Service
public class StudentServiceImpl implements IStudentService{
    @Autowired
    private IStudentDao dao;
    @Autowired
    private ITbClassDao tbclassDao;
    @Override
    public PageInfo<Student> queryAllStudentMessage(Integer pageNum, Integer pageSize) throws Exception {
        PageHelper.startPage(pageNum, pageSize);
        List<Student> list = dao.queryAllStudentMessage();
        PageInfo<Student> info = new PageInfo<>(list);
```

```java
        return info;
    }
    @Override
    public List<Student> dimQureyStudentByClassAndSidAndName(String clsid, Integer stuid, String sname)
            throws Exception {
        return dao.dimQureyStudentByClassAndSidAndName(clsid, stuid, sname);
    }
    @Override
    public Boolean insertStudent(Student student) throws Exception {
        student.setStuid(IDUtil.getIDUtil());
        student.setStatus(1);
        Boolean flag = dao.insertStudent(student) > 0;
        if(flag) {
            Integer count = tbclassDao.queryStuCountById(student.getClsid());
            tbclassDao.updateClassPersonNumber(student.getClsid(), (count+1));
        }
        return flag;
    }
    @Override
    public Boolean updateStudentMessage(Student student) throws Exception {
        Student flagClass = dao.queryStudentById(student.getStuid());
        if(student.getClsid() != flagClass.getClsid()) {
            //如果修改后的班级和原来的不一样
            System.out.println("班级修改了");
            //查询原来班级的人数总和
            Integer nowCount = tbclassDao.queryStuCountById(flagClass.getClsid());
            //查看修改后班级的人数总和
            Integer updateCount = tbclassDao.queryStuCountById(student.getClsid());
            //原来班级人数-1
            tbclassDao.updateClassPersonNumber(flagClass.getClsid(), (nowCount-1));
            //修改后的班级+1
            tbclassDao.updateClassPersonNumber(student.getClsid(), (updateCount+1));
        }
        return dao.updateStudentMessage(student) > 0;
    }
    @Override
    public Boolean expelStudent(Integer status, String stuid) throws Exception {
        return dao.expelStudent(status, stuid) > 0;
    }
    @Override
    public PageInfo<Student> queryAllClassAndStudent(Integer pageNum, Integer pageSize, Map<String, Object> map)
            throws Exception {
        PageHelper.startPage(pageNum, pageSize);
        List<Student> list = dao.queryAllClassAndStudent(map);
        PageInfo<Student> info = new PageInfo<>(list);
        return info;
    }
```

```java
    @Override
public Student queryStudentById(String stuid) throws Exception {
    return dao.queryStudentById(stuid);
}
    @Override
public PageInfo<Student> queryByStudentByClsid(Integer clsid, IntegerpageNum, IntegerpageSize) throws Exception {
    PageHelper.startPage(pageNum, pageSize);
    List<Student> list = dao.queryByStudentByClsid(clsid);
    PageInfo<Student> info = new PageInfo<>(list);
    return info;
}
}
```

活动服务层接口(ITbActivityService)以及对应实现类(TbActivityServiceImpl)如下所示。

```java
package com.test.service;
import java.util.List;
import com.github.pagehelper.PageInfo;
import com.test.bean.TbActivity;
/**
 * 服务层活动表
 */
public interface ITbActivityService {
// ---------------------------------------------------------
    // 班主任操作
/**
 * 创建活动
 *
 * @param activity
 */
public Boolean createActivity(TbActivity activity) throws Exception;
/**
 * 修改活动
 *
 * @param activity
 */
public Boolean updateActivity(TbActivity activity) throws Exception;
/**
 * 分页展示活动
 */
public PageInfo<TbActivity> queryAllActivity(Integer pageNum, IntegerpageSize) throws Exception;
/**
 * 单击活动显示活动详情
 *
 * @param id
 */
public TbActivity queryActivityMessage(Integer id) throws Exception;
}
```

```java
package com.test.service.impl;
import java.util.List;
import org.springframework.beans.factory.annotation.Autowired;
import org.springframework.context.ApplicationContext;
import org.springframework.context.support.ClassPathXmlApplicationContext;
import org.springframework.stereotype.Service;
import com.github.pagehelper.PageHelper;
import com.github.pagehelper.PageInfo;
import com.test.bean.TbActivity;
import com.test.dao.ITbActivityDao;
import com.test.service.ITbActivityService;
import com.test.util.IDUtil;
@Service
public class TbActivityServiceImpl implements ITbActivityService{
    @Autowired
    private ITbActivityDao dao;
    @Override
    public Boolean createActivity(TbActivity activity) throws Exception {
        //插入时间
        activity.setStarttime(IDUtil.getTimeUtil());
        return dao.createActivity(activity)>0;
    }
    @Override
    public Boolean updateActivity(TbActivity activity) throws Exception {
        return dao.updateActivity(activity)>0;
    }
    @Override
    public PageInfo<TbActivity> queryAllActivity(Integer pageNum,Integer pageSize) throws Exception {
        PageHelper.startPage(pageNum, pageSize);
        List<TbActivity> list = dao.queryAllActivity();
        PageInfo<TbActivity> info = new PageInfo<>(list);
        return info;
    }
    @Override
    public TbActivity queryActivityMessage(Integer id) throws Exception {
        return dao.queryActivityMessage(id);
    }
    public static void main(String[] args) {
        ApplicationContext ac = new ClassPathXmlApplicationContext("applicationContext.xml");
        TbActivityServiceImpl dao = ac.getBean(TbActivityServiceImpl.class);
        try {
            PageInfo<TbActivity> list = dao.queryAllActivity(1, 5);
            System.out.println(list);
        } catch (Exception e) {
            // TODO Auto-generated catch block
            e.printStackTrace();
        }
    }
}
```

班级服务层接口(ITbClassService)以及对应实现类(TbClassServiceImpl)如下所示。

```java
package com.test.service;
import java.util.List;
import com.github.pagehelper.PageInfo;
import com.test.bean.TbClass;
/**
 * 服务层接口班级管理
 */
public interface ITbClassService{
/**
 * 分页展示班级信息
 */
public PageInfo<TbClass>dimQueryTbClassMessage(Integer pageNum,IntegerpageSize) throws Exception;
/**
 * 开设班级
 * @param tbclass
 */
public Boolean insertTbClass(TbClass tbclass) throws Exception;
/**
 * 修改班级班主任
 * @param teacher
 * @param cid
 */
public Boolean updateTbClass(Integer teacher,Integercid) throws Exception;
/**
 * 通过班级名称查询班级
 */
public TbClass queryByClassName() throws Exception;
/**
 * 查询所有班级名称信息
 */
public List<TbClass> queryAllClassMessage() throws Exception;
/**
 * 修改班级资料
 * @param tbclass
 */
public Boolean updateClassMessage(TbClass tbclass) throws Exception;

/**
 * 通过班级id查询修改班级信息
 * @param cid
 */
public TbClass queryClassById(Integer cid) throws Exception;
}
package com.test.service.impl;
import java.util.List;
import org.springframework.beans.factory.annotation.Autowired;
```

```java
import org.springframework.stereotype.Service;
import com.github.pagehelper.PageHelper;
import com.github.pagehelper.PageInfo;
import com.test.bean.TbClass;
import com.test.dao.ITbClassDao;
import com.test.service.ITbClassService;
@Service
public class TbClassServiceImpl implements ITbClassService{
    @Autowired
    private ITbClassDao dao;
    @Override
    public PageInfo<TbClass> dimQueryTbClassMessage(Integer pageNum, Integer pageSize) throws Exception{
        PageHelper.startPage(pageNum, pageSize);
        List<TbClass> list = dao.dimQueryTbClassMessage();
        PageInfo<TbClass> info = new PageInfo<>(list);
        return info;
    }
    @Override
    public Boolean insertTbClass(TbClass tbclass) throws Exception{
        tbclass.setStucount(0);
        return dao.insertTbClass(tbclass)>0;
    }
    @Override
    public Boolean updateTbClass(Integer teacher, Integer cid) throws Exception{
        return dao.updateTbClass(teacher, cid)>0;
    }
    @Override
    public TbClass queryByClassName() throws Exception{
        return dao.queryByClassName();
    }
    @Override
    public List<TbClass> queryAllClassMessage() throws Exception{
        return dao.dimQueryTbClassMessage();
    }
    @Override
    public Boolean updateClassMessage(TbClass tbclass) throws Exception{
        return dao.updateClassMessage(tbclass)>0;
    }
    @Override
    public TbClass queryClassById(Integer cid) throws Exception{
        return dao.queryClassById(cid);
    }
}
```

资料服务层接口(ITbResourceService)以及对应实现类(TbResourceServiceImpl)如下所示。

```java
package com.test.service;
import java.util.List;
```

```java
import com.github.pagehelper.PageInfo;
import com.test.bean.TbResource;
/**
 * 服务层功能*
 */
public interface ITbResourceService{
// -----------------------------------------------------------
    // 班主任操作
/**
 上传班级资料
    * @param reource
 */
public Boolean uploadTbClassData(TbResource reource) throws Exception;
/**
    * 资料展示
 */
public PageInfo<TbResource>queryResource(Integer pageNum,IntegerpageSize) throws Exception;
/**
    * 通过班级id查询班级资料
    * @param clsid
 */
public List<TbResource>queryByResourceByClsid(Integer clsid) throws Exception;;
/**
    * 修改班级资料
* @param resource
 */
public Boolean updateClassData(TbResource resource) throws Exception;
/**
    * 通过id查询资料
    * @param id
 */
public TbResource queryResourceById(Integer id) throws Exception;
}
package com.test.service.impl;
import java.util.List;
import org.springframework.beans.factory.annotation.Autowired;
import org.springframework.stereotype.Service;
import com.github.pagehelper.PageHelper;
import com.github.pagehelper.PageInfo;
import com.test.bean.TbResource;
import com.test.dao.ITbResourceDao;
import com.test.service.ITbResourceService;
@Service
public class TbResourceServiceImplimplements ITbResourceService{
@Autowired
private ITbResourceDaodao;
@Override
```

```java
public Boolean uploadTbClassData(TbResource reource) throws Exception{
    return dao.uploadTbClassData(reource)>0;
}
@Override
public PageInfo<TbResource>queryResource(Integer pageNum, Integer pageSize) throws Exception{
    PageHelper.startPage(pageNum,pageSize);
    List<TbResource>list = dao.queryResource();
    PageInfo<TbResource>info = new PageInfo<>(list);
    return info;
}
@Override
public List<TbResource>queryByResourceByClsid(Integer clsid) throws Exception{
    return dao.queryByResourceByClsid(clsid);
}
@Override
public Boolean updateClassData(TbResource resource) throws Exception{
    return dao.updateClassData(resource)>0;
}
@Override
public TbResource queryResourceById(Integer id) throws Exception{
    return dao.queryResourceById(id);
}
}
```

教师服务层接口(ITeacherService)以及对应实现类(TeacherServiceImpl)如下所示。

```java
package com.test.service;
import java.util.List;
import org.apache.ibatis.annotations.Param;
import com.github.pagehelper.PageInfo;
import com.test.bean.Teacher;
/**
 * 业务层教师管理
 */
public interface ITeacherService{

    /**
     * 分页展示老师信息
     */
    public PageInfo<Teacher>queryAllTeacherMessage(Integer pageNum, Integer pageSize) throws Exception;
    /**
     * 添加老师信息
     * @param teacher
     */
    public Boolean insertTeacher(Teacher teacher) throws Exception;
    /**
     * 开除老师(不删除仅修改状态)/恢复开除老师状态
     *
```

```java
     * @param tid
 * @param status
 */
public Boolean expelTeacher(Integer tid, Integer status) throws Exception;
/**
     * 禁止登录/解除禁止登录
     * @param tid
 * @param islogin
 */
public Boolean forbidTeacherLogin(Integer tid, Integer islogin) throws Exception
/**
     * 通过班主任名称查询是否存在
 */
public Teacher queryByTeacherName() throws Exception;
    // 班主任操作
/**
     * 班主任登录
     * @param teacher
 */
public Teacher teacherLogin(Teacher teacher) throws Exception;
/**
     * 班主任修改密码
     * @param tid 班主任编号
 * @param password 密码
 */
public Boolean teacherUpdatePassword(@Param("id") Integer tid, @Param("password") Integer password)
    throws Exception;
/**
     * 班主任修改个人资料
     * @param teacher
 */
public Boolean teacherUpdatePersonalData(Teacher teacher) throws Exception;
    //管理员
/**
     * 查询所有老师信息
 */
public List<Teacher> queryAllTeacher() throws Exception;
/**
     * 管理员操作
 * 分页展示班级信息两表联合
 * @param tid
 */
public Teacher queryById(Integer tid) throws Exception;
}
package com.test.service.impl;
import java.util.List;
import org.springframework.beans.factory.annotation.Autowired;
import org.springframework.stereotype.Service;
```

```java
import com.github.pagehelper.PageHelper;
import com.github.pagehelper.PageInfo;
import com.test.bean.Teacher;
import com.test.dao.ITeacherDao;
import com.test.service.ITeacherService;
@Service
public class TeacherServiceImpl implements ITeacherService{
    @Autowired
    private ITeacherDao dao;
    @Override
    public PageInfo<Teacher> queryAllTeacherMessage(Integer pageNum, Integer pageSize) throws Exception {
        PageHelper.startPage(pageNum, pageSize);
        List<Teacher> list = dao.queryAllTeacherMessage();
        PageInfo<Teacher> info = new PageInfo<>(list);
        return info;
    }
    @Override
    public Boolean insertTeacher(Teacher teacher) throws Exception {
        return dao.insertTeacher(teacher)>0;
    }
    @Override
    public Boolean expelTeacher(Integer tid, Integer status) throws Exception {
        return dao.expelTeacher(tid, status)>0;
    }
    @Override
    public Boolean forbidTeacherLogin(Integer tid, Integer islogin) throws Exception {
        return dao.forbidTeacherLogin(tid, islogin)>0;
    }
    @Override
    public Teacher queryByTeacherName() throws Exception {
        return dao.queryByTeacherName();
    }
    @Override
    public Teacher teacherLogin(Teacher teacher) throws Exception {
        return dao.teacherLogin(teacher);
    }
    @Override
    public Boolean teacherUpdatePassword(Integer tid, Integer password) throws Exception {
        return dao.teacherUpdatePassword(tid, password)>0;
    }
    @Override
    public Boolean teacherUpdatePersonalData(Teacher teacher) throws Exception {
        return dao.teacherUpdatePersonalData(teacher)>0;
    }
    @Override
    public List<Teacher> queryAllTeacher() throws Exception {
        return dao.queryAllTeacherMessage();
    }
```

```
@ Override
public Teacher queryById(Integer tid) throws Exception {
return dao.queryById(tid);
    }
}
```

15.8　创建页面视图(view)

登录视图效果如图 15-7 所示。

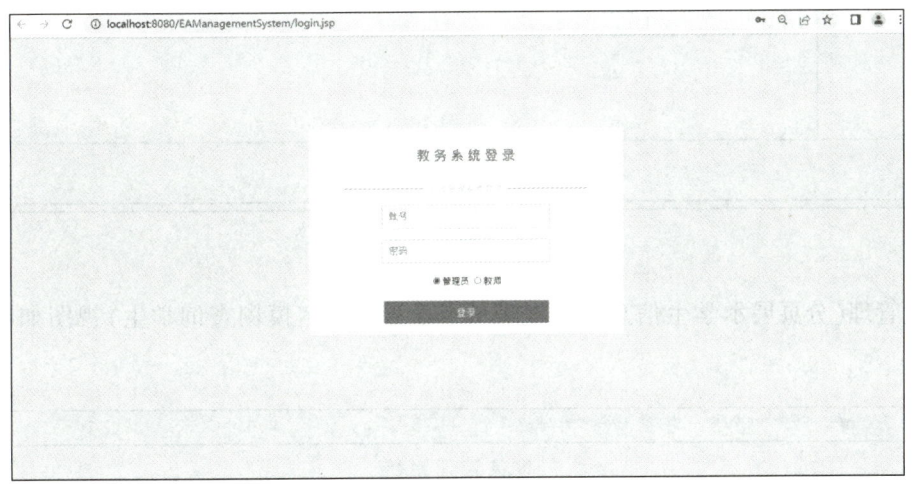

图 15-7　登录视图

管理员对象进入系统后,修改本人资料视图如图 15-8 所示。

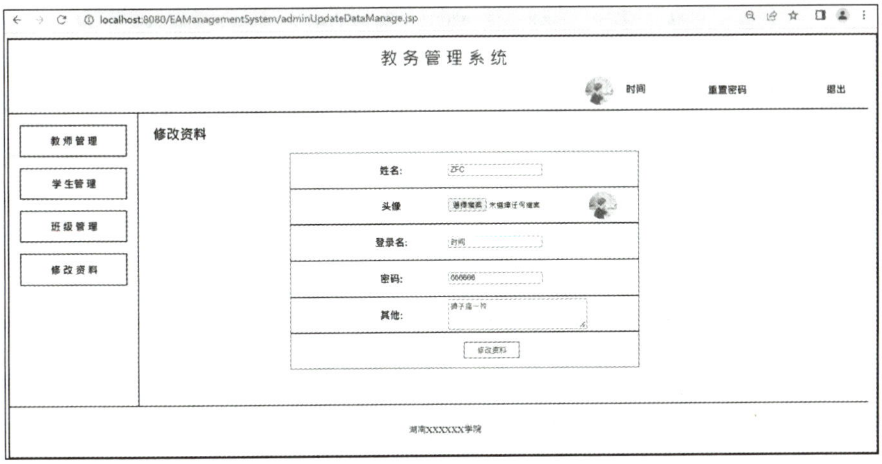

图 15-8　管理员修改本人资料视图

教师管理[分页展示老师信息，添加老师信息，开除老师(不删除仅修改状态)]，禁止登录视图如图 15-9 所示。

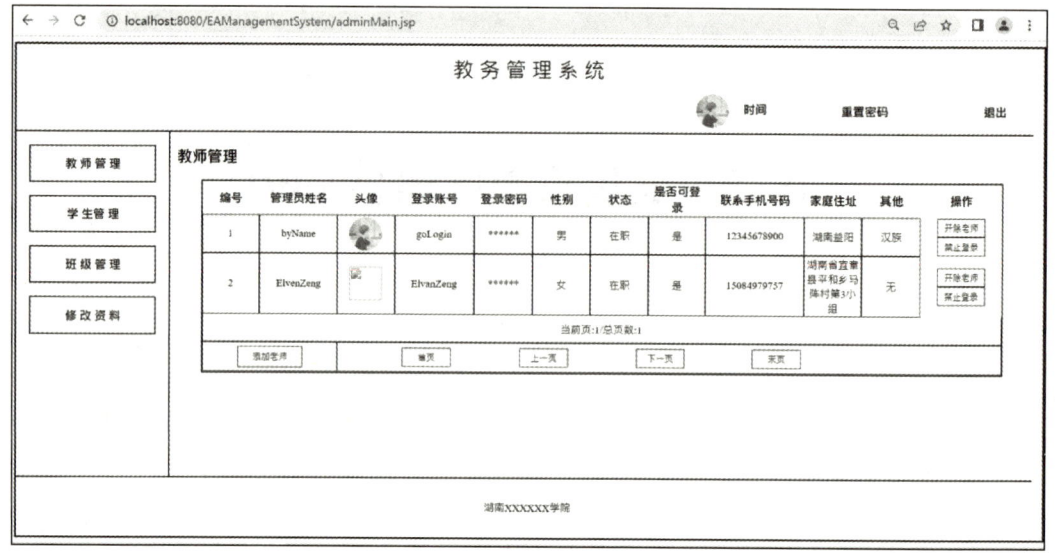

图 15-9　教师管理视图

学生管理(分页展示学生信息，根据班级、学号、姓名模糊查询学生)视图如图 15-10 所示。

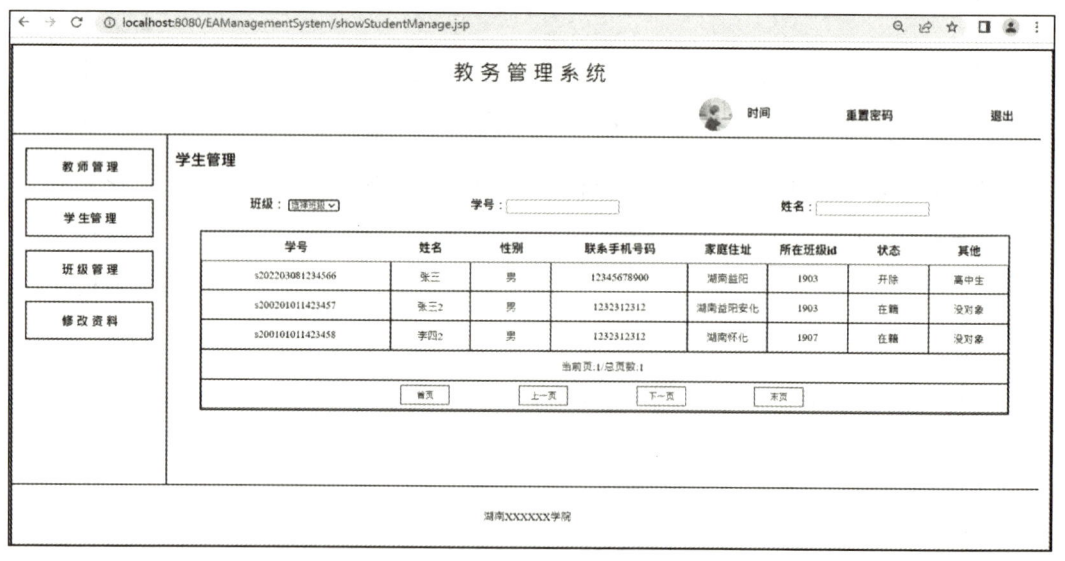

图 15-10　学生管理视图

班级管理(分页展示班级信息，开设班级，修改班级班主任)视图如图 15-11 所示。

第 15 章　教务管理系统

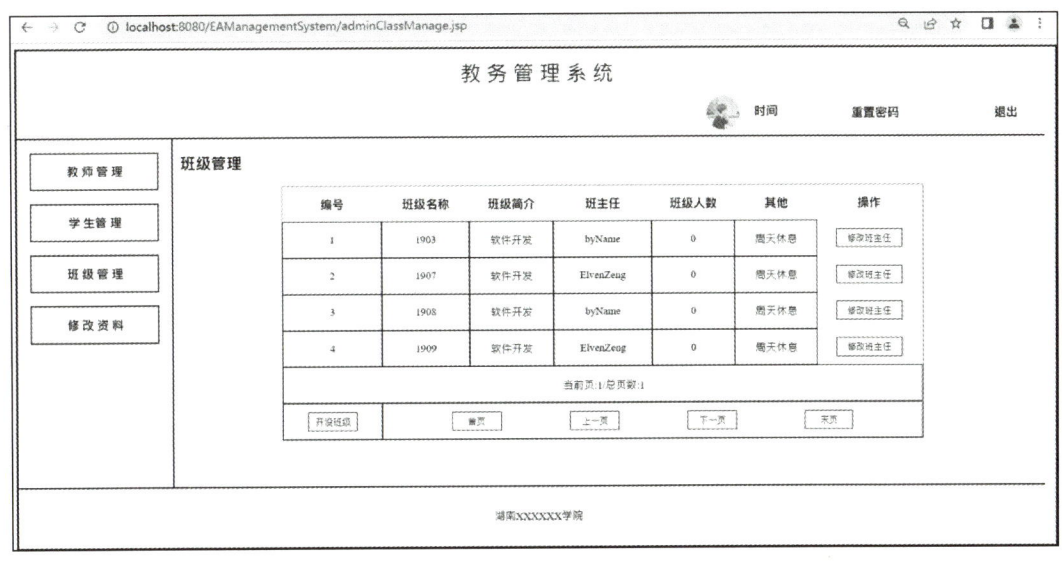

图 15-11　班级管理视图

班主任对象进入系统，修改本人资料视图如图 15-12 所示。

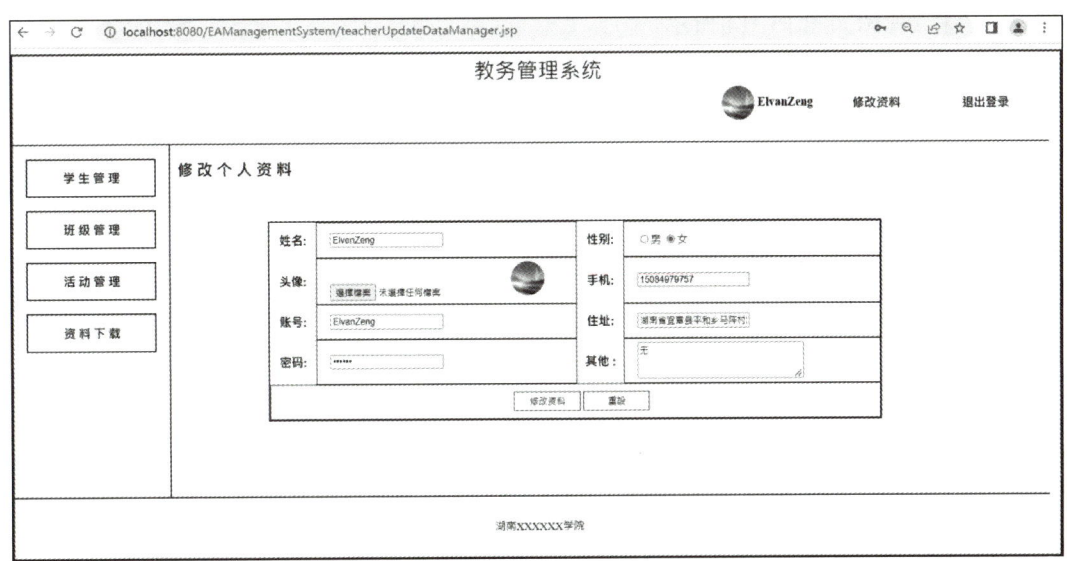

图 15-12　班主任修改本人资料视图

学生管理[分页查询所带班级学生信息(根据姓名、学号、班级、状态模糊查询)，添加学生，修改学生信息，开除学生(不删除仅修改状态)]视图如图 15-13 所示。

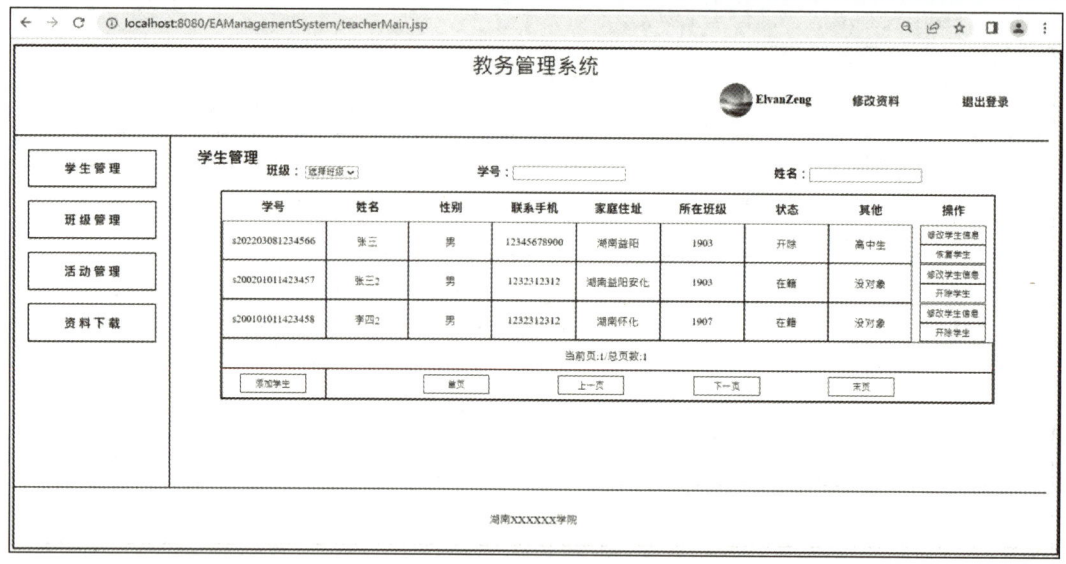

图 15-13　学生管理视图

班级管理(分页展示所有班级，单击班级在下方分页显示班级学生信息，修改班级资料，上传班级资料)视图如图 15-14 所示。

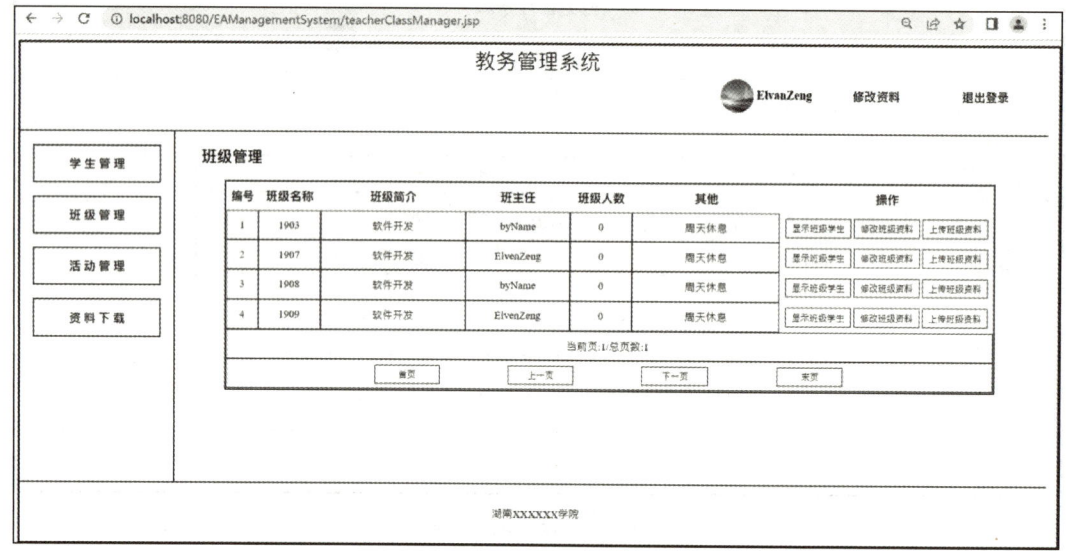

图 15-14　班级管理视图

活动管理(创建活动，修改活动，分页展示活动，单击活动显示活动详情)视图如图 15-15 所示。

第 15 章 教务管理系统

图 15-15　活动管理视图

班级资料下载(资料展示资料下载)视图如图 15-16 所示。

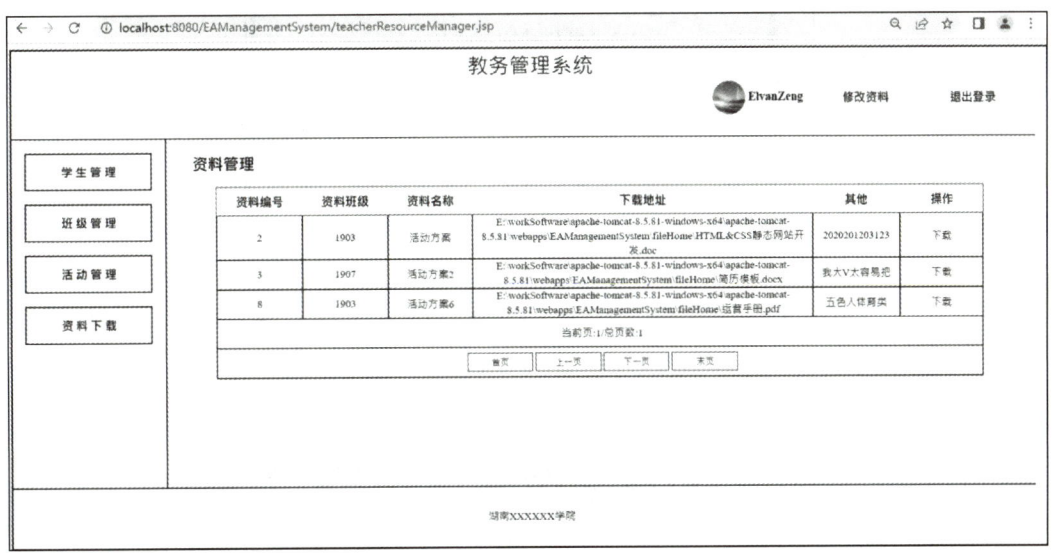

图 15-16　资料管理视图

webapp 目录结构如图 15-17 所示。

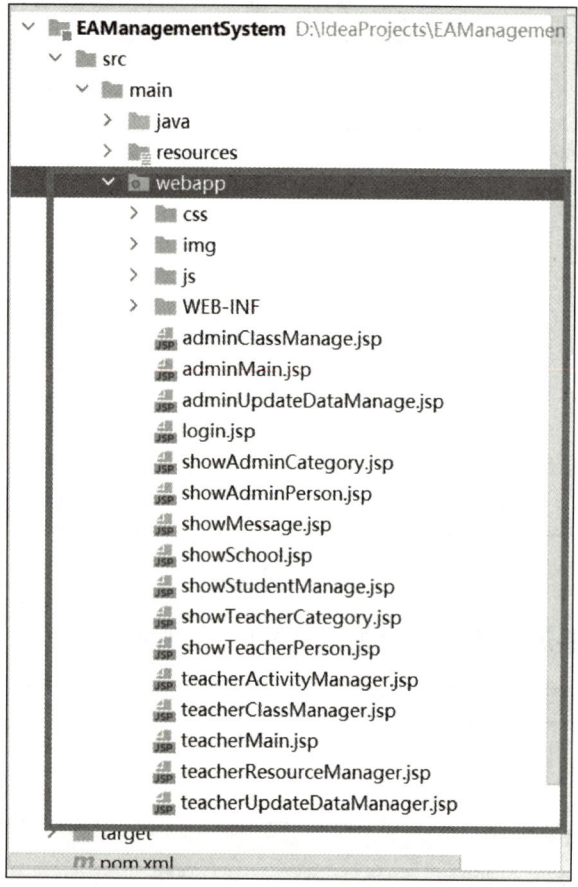

图 15-17 目录结构图

图 15-17 所示 webapp 源码资源（css、js、jsp）请到如下地址下载：https://gitee.com/ElvanZeng/booksource.git。

15.9 创建控制层（controller）

管理员控制层（AdminController）如下所示。

```
package com.test.controller;
import java.io.File;
import java.io.IOException;
import java.util.HashMap;
import java.util.Map;
import javax.servlet.http.HttpServletRequest;
import javax.servlet.http.HttpSession;
```

```java
import org.springframework.beans.factory.annotation.Autowired;
import org.springframework.stereotype.Controller;
import org.springframework.ui.Model;
import org.springframework.web.bind.annotation.RequestMapping;
import org.springframework.web.bind.annotation.RequestParam;
import org.springframework.web.bind.annotation.ResponseBody;
import org.springframework.web.multipart.MultipartFile;
import com.test.bean.Admin;
import com.test.bean.Teacher;
import com.test.service.IAdminService;
@Controller
@RequestMapping("admin")
public class AdminController{
@Autowired
private IAdminService service;
@RequestMapping("adminLogin")
@ResponseBody
public Map<String, Object>adminLogin(Admin admin, HttpSession session){
Map<String, Object>map = new HashMap<String, Object>();
try{
Admin adminMg= service.adminLogin(admin);
if(adminMg!=null){
session.setAttribute("admin", adminMg);
map.put("sAnde","管理员登录成功");
return map;
}
}catch(Exception e){
// TODO Auto-generated catch block
e.printStackTrace();
}
map.put("sAnde","账号不存在,或密码输入错误");
return map;
}
@RequestMapping("exitLogin")
public String exitLogin(HttpSession session){
//销毁管理员
Admin admin= (Admin)session.getAttribute("admin");
if(admin!=null){
session.removeAttribute("admin");
}
//销毁老师
Teacher teacher= (Teacher)session.getAttribute("teacher");
if(teacher!=null){
session.removeAttribute("teacher");
}
return "redirect:/login.jsp";
}
```

```java
@RequestMapping("resetPassword")
public String resetPassword(HttpSession session){
    //重置密码
    Admin admin = (Admin)session.getAttribute("admin");
    if(admin!=null){
        try{
            service.adminResertPS(admin.getAid());
            session.removeAttribute("admin");
        }catch (Exception e){
            // TODO Auto-generated catch block
            e.printStackTrace();
        }
    }
    return "redirect:/login.jsp";
}

@RequestMapping("adminUpdateByThis")
public String adminUpdateByThis(Admin admin,@RequestParam("myFile")MultipartFile myFile,
        HttpServletRequest request,
        Model model,HttpSession session){
    Boolean flag = false;
    try{
        //1.设置上传路径
        String storPath = request.getServletContext().getRealPath("img");
        //2.判断该位置存在不存在
        File file = new File(storPath,myFile.getOriginalFilename());
        if(!file.getParentFile().exists()){//如果父路径不存在
            file.mkdirs();//则创建目录
        }
        //3.将文件上传到上传路径
        try{
            myFile.transferTo(file);
        }catch (IllegalStateException | IOException e){
            // TODO Auto-generated catch block
            e.printStackTrace();
        }
        admin.setPhoto(myFile.getOriginalFilename());
        flag = service.adminUpdateByThis(admin);
        if(flag){
            //修改成功,重新存session
            Admin adminSession = service.adminLogin(admin);
            session.setAttribute("admin",adminSession);
        }
        //前端显示是否修改成功
        String adminUpdateFlag = flag?"修改个人信息成功":"修改个人信息失败";
        model.addAttribute("adminUpdateFlag",adminUpdateFlag);
    }catch (Exception e){
        // TODO Auto-generated catch block
```

```
            e.printStackTrace();
        }
        return "/adminMain.jsp";
    }
    @RequestMapping("queryAdminById")
    @ResponseBody
    public Admin queryAdminById(HttpSession session){
        Admin sessionAdmin = (Admin)session.getAttribute("admin");
        Admin admin = null;
        try{
            admin = service.queryAdminById(sessionAdmin.getAid());
        }catch(Exception e){
            //TODO Auto-generated catch block
            e.printStackTrace();
        }
        return admin;
    }
}
```

学生控制层(StudentController)如下所示。

```
package com.test.controller;
import java.util.HashMap;
import java.util.Map;
import javax.servlet.http.HttpSession;
import org.springframework.beans.factory.annotation.Autowired;
import org.springframework.stereotype.Controller;
import org.springframework.ui.Model;
import org.springframework.web.bind.annotation.RequestMapping;
import org.springframework.web.bind.annotation.ResponseBody;
import com.github.pagehelper.PageInfo;
import com.test.bean.Student;
import com.test.service.IStudentService;
@Controller
@RequestMapping("/student")
public class StudentController{
    @Autowired
    private IStudentService service;
    @RequestMapping("/queryStudentAll")
    @ResponseBody
    public PageInfo<Student> queryStudentAll(Integer pageNum,Integer pageSize,
        String clsid,Integer stuid,String name){
        PageInfo<Student> info = null;
        try{
            Map<String,Object> map = new HashMap<>();
            map.put("clsid",clsid);
            map.put("stuid",stuid);
```

```java
            map.put("sname", sname);
            info = service.queryAllClassAndStudent(pageNum, pageSize, map);
        } catch (Exception e) {
            // TODO Auto-generated catch block
            e.printStackTrace();
        }
        return info;
    }
    @RequestMapping("/insertStudent")
    public String insertStudent(Student student, Model model, HttpSession session) {
        try {
            Boolean flag = service.insertStudent(student);
            String insertFlag = flag?"添加成功":"添加失败";
            System.out.println("-------------"+insertFlag);
            model.addAttribute("insertFlag", insertFlag);
        } catch (Exception e) {
            // TODO Auto-generated catch block
            e.printStackTrace();
        }
        return "/teacherMain.jsp";
    }
    @RequestMapping("/queryStudentById")
    @ResponseBody
    public Student queryStudentById(String stuid) {
        Student student = null;
        try {
            student = service.queryStudentById(stuid);
        } catch (Exception e) {
            e.printStackTrace();
        }
        return student;
    }
    @RequestMapping("/updateStudentMessage")
    public String updateStudentMessage(Student student, Model model) {
        try {
            Boolean flag = service.updateStudentMessage(student);
            String stuUpdateFlag = flag?"修改成功":"修改失败";
            model.addAttribute("stuUpdateFlag", stuUpdateFlag);
        } catch (Exception e) {
            e.printStackTrace();
        }
        return "/teacherMain.jsp";
    }

    @RequestMapping("/expelStudent")
    @ResponseBody
    public Boolean expelStudent(String stuid, Integer status) {
```

```java
        Boolean flag = false;
try {
        flag = service.expelStudent(status,stuid);
    } catch (Exception e) {
// TODO Auto-generated catch block
e.printStackTrace();
    }
return flag;
    }
    @RequestMapping("queryStudentMessageByClsid")
    @ResponseBody
public PageInfo<Student> queryStudentMessageByClsid(Integer cid,IntegerpageNum,IntegerpageSize){
PageInfo<Student> info = null;
try {
        info = service.queryByStudentByClsid(cid,pageNum,pageSize);
    } catch (Exception e) {
e.printStackTrace();
    }
return info;
    }
}
```

活动控制层(ActivityController)如下所示。

```java
package com.test.controller;
import org.springframework.beans.factory.annotation.Autowired;
import org.springframework.stereotype.Controller;
import org.springframework.ui.Model;
import org.springframework.web.bind.annotation.RequestMapping;
import org.springframework.web.bind.annotation.ResponseBody;
import com.github.pagehelper.PageInfo;
import com.test.bean.TbActivity;
import com.test.service.impl.TbActivityServiceImpl;
@Controller
@RequestMapping("activity")
public class ActivityController{
@Autowired
private TbActivityServiceImplservice;
@RequestMapping("queryAllActivity")
@ResponseBody
public PageInfo<TbActivity> queryAllActivity(Integer pageNum,IntegerpageSize){
PageInfo<TbActivity> info = null;
try {
        info = service.queryAllActivity(pageNum,pageSize);
    } catch (Exception e) {
e.printStackTrace();
    }
return info;
```

```java
        }
    @RequestMapping("createActivity")
    public String createActivity(TbActivity activity,Model model){
        try{
            Boolean flag = service.createActivity(activity);
            String createFlag= flag?"操作成功":"创建失败";
            model.addAttribute("createFlag",createFlag);
        }catch (Exception e){
            e.printStackTrace();
        }
        return "/teacherActivityManager.jsp";
    }
    @RequestMapping("queryActivityById")
    @ResponseBody
    public TbActivity queryActivityById(Integer id){
        TbActivity activity = null;
        try{
            activity = service.queryActivityMessage(id);
        }catch (Exception e){
            e.printStackTrace();
        }
        return activity;
    }
    @RequestMapping("updateActivity")
    public String updateActivity(TbActivity activity,Model model){
        try{
            Boolean flag = service.updateActivity(activity);
            String updateFlag= flag?"操作成功":"修改失败";
            model.addAttribute("createFlag",updateFlag);
        }catch (Exception e){
            e.printStackTrace();
        }
        return "/teacherActivityManager.jsp";
    }
    @RequestMapping("queryActivityMessage")
    @ResponseBody
    public TbActivity queryActivityMessage(Integer id,Model model){
        TbActivity activity = null;
        try{
            activity = service.queryActivityMessage(id);
            model.addAttribute("activity",activity);
        }catch (Exception e){
            e.printStackTrace();
        }
        return activity;
    }
}
```

班级控制层(ClassController)如下所示。

```java
package com.test.controller;
import java.util.ArrayList;
import java.util.HashMap;
import java.util.List;
import java.util.Map;
import org.springframework.beans.factory.annotation.Autowired;
import org.springframework.stereotype.Controller;
import org.springframework.ui.Model;
import org.springframework.web.bind.annotation.RequestMapping;
import org.springframework.web.bind.annotation.ResponseBody;
import com.github.pagehelper.PageInfo;
import com.test.bean.TbClass;
import com.test.service.ITbClassService;
@Controller
@RequestMapping("class")
public class ClassController{
    @Autowired
    private ITbClassService service;
    @RequestMapping("queryAllClassMessage")
    @ResponseBody
    public Map<String,Object>queryAllClassMessage(){
    Map<String,Object>map = new HashMap<String,Object>();
    List<TbClass> list = new ArrayList<TbClass>();
    try{
            list = service.queryAllClassMessage();
    map.put("list",list);
        }catch(Exception e){
    e.printStackTrace();
        }
    return map;
      }
    @RequestMapping("queryAllTeacherMessag")
    @ResponseBody
    public PageInfo<TbClass>queryAllTeacherMessag(Integer pageNum,Integer pageSize){
    PageInfo<TbClass> info = null;
    try{
            info = service.dimQueryTbClassMessage(pageNum,pageSize);
        }catch(Exception e){
    e.printStackTrace();
        }
    return info;
      }
    @RequestMapping("updateTeacherById")
        @ResponseBody
    public Boolean updateTeacherById(Integer teacher,Integer cid){
        Boolean flag = false;
```

```java
        try {
        //老师id,班级id
        flag = service.updateTbClass(teacher,cid);
            } catch (Exception e) {
        e.printStackTrace();
            }
        return flag;
        }
    @RequestMapping("insertTbClass")
    @ResponseBody
    public Boolean insertTbClass(TbClass tbclass,Model model) {
        Boolean flag = false;
        try {
            flag = service.insertTbClass(tbclass);
            String insertTbClassFlag = flag?"开设成功":"开设失败";
    model.addAttribute("insertTbClassFlag", insertTbClassFlag);
            } catch (Exception e) {
        e.printStackTrace();
            }
        return flag;
        }
    @RequestMapping("queryClassById")
    @ResponseBody
    public TbClass queryClassById(Integer cid) {
        TbClass tbclass = null;
        try {
            tbclass = service.queryClassById(cid);
            } catch (Exception e) {
        e.printStackTrace();
            }
        return tbclass;
        }
    @RequestMapping("updateClassById")
    public String updateClassById(TbClass tbclass,Model model) {
        try {
            Boolean flag = service.updateClassMessage(tbclass);
            String updateTbClassFlag = flag?"操作成功":"修改失败";
    model.addAttribute("insertTbClassFlag", updateTbClassFlag);
            } catch (Exception e) {
        e.printStackTrace();
            }
        return "/teacherClassManager.jsp";
        }
    }
```

第 15 章　教务管理系统

资料控制层（ResourceController）如下所示。

```java
package com.test.controller;
import java.io.File;
import java.io.IOException;
import java.util.HashMap;
import java.util.List;
import java.util.Map;
import javax.servlet.http.HttpServletRequest;
import org.apache.commons.io.FileUtils;
import org.springframework.beans.factory.annotation.Autowired;
import org.springframework.http.HttpHeaders;
import org.springframework.http.HttpStatus;
import org.springframework.http.MediaType;
import org.springframework.http.ResponseEntity;
import org.springframework.stereotype.Controller;
import org.springframework.ui.Model;
import org.springframework.web.bind.annotation.RequestMapping;
import org.springframework.web.bind.annotation.RequestParam;
import org.springframework.web.bind.annotation.ResponseBody;
import org.springframework.web.multipart.MultipartFile;
import com.github.pagehelper.PageInfo;
import com.test.bean.TbResource;
import com.test.service.ITbResourceService;
@Controller
@RequestMapping("resource")
public class ResourceController {
    @Autowired
    private ITbResourceService service;
    @RequestMapping("queryByResourceByClsid")
    @ResponseBody
    public Map<String, Object>queryByResourceByClsid(Integer clsid) {
        Map<String, Object> map = new HashMap<String, Object>();
        try {
            List<TbResource>classResource = service.queryByResourceByClsid(clsid);
            map.put("classResource", classResource);
        } catch (Exception e) {
            e.printStackTrace();
        }
        return map;
    }
    @RequestMapping("updateClassData")
    public String updateClassData(TbResource resource, Model model, @RequestParam("myFile") MultipartFilemyFile,
HttpServletRequest request) {
        // 1.设置上传路径
        String storPath = request.getServletContext().getRealPath("fileHome");
        // 2.判断该位置存在不存在
```

```java
            File file = new File(storPath, myFile.getOriginalFilename());
            String url = file.toString();
resource.setUrl(url);
            if(!file.getParentFile().exists()){// 如果父路径不存在
file.mkdirs();// 则创建目录
            }
            try{
                // 3.将文件上传到上传路径
myFile.transferTo(file);
                Boolean flag = service.updateClassData(resource);
                String resourceUpdate = flag ? "操作成功" : "修改失败";
model.addAttribute("resourceUpdate", resourceUpdate);
            } catch (Exception e){
e.printStackTrace();
            }
            return "/teacherClassManager.jsp";
        }
@RequestMapping("queryResourceById")
    @ResponseBody
public TbResource queryResourceById(Integer id){
TbResource resource = null;
try{
            resource = service.queryResourceById(id);
        } catch (Exception e){
e.printStackTrace();
        }
return resource;
    }
    @RequestMapping("uploadTbClassData")
public String uploadTbClassData(TbResource resource, Model model, @RequestParam("myFile") MultipartFile myFile,
HttpServletRequest request){
// 1.设置上传路径
String storPath = request.getServletContext().getRealPath("fileHome");
// 2.判断该位置存在不存在
File file = new File(storPath, myFile.getOriginalFilename());
        String url = file.getAbsolutePath();//获得全路径
resource.setUrl(url);
if(!file.getParentFile().exists()){// 如果父路径不存在
file.mkdirs();// 则创建目录
        }
try{
// 3.将文件上传到上传路径
myFile.transferTo(file);
Boolean flag = service.uploadTbClassData(resource);
String resourceUpdate = flag ? "操作成功" : "修改失败";
model.addAttribute("resourceUpdate", resourceUpdate);
```

```java
        }catch(Exception e){
e.printStackTrace();
    }
return "/teacherClassManager.jsp";
    }
@RequestMapping("/queryAllResource")
@ResponseBody
public PageInfo<TbResource> queryAllResource(Integer pageNum, Integer pageSize){
PageInfo<TbResource> info = null;
try{
        info = service.queryResource(pageNum, pageSize);
    }catch(Exception e){
e.printStackTrace();
    }
return info;
    }
    @RequestMapping("/downloadResource")
public ResponseEntity<byte[]> downloadResource(@RequestParam("fileName") String fileName,
HttpServletRequest request) throws IOException{
//获得下载的路径
String stroeHome = request.getServletContext().getRealPath("fileHome");
//设置字符编码
String fileNameStr = new String(fileName.getBytes("UTF-8"), "ISO-8859-1");
        File file = new File(stroeHome + File.separator + fileName);
//设置请求体
HttpHeaders headers = new HttpHeaders();
headers.setContentDispositionFormData("attachment", fileNameStr);
headers.setContentType(MediaType.APPLICATION_OCTET_STREAM);
return new ResponseEntity<byte[]>(FileUtils.readFileToByteArray(file), headers, HttpStatus.CREATED);
    }
}
```

教师控制层(TeacherController)如下所示。

```java
package com.test.controller;
import java.io.File;
import java.io.IOException;
import java.util.HashMap;
import java.util.List;
import java.util.Map;
import javax.servlet.http.HttpServletRequest;
import javax.servlet.http.HttpSession;
import org.springframework.beans.factory.annotation.Autowired;
import org.springframework.stereotype.Controller;
import org.springframework.ui.Model;
import org.springframework.web.bind.annotation.RequestMapping;
import org.springframework.web.bind.annotation.RequestParam;
```

```java
import org.springframework.web.bind.annotation.ResponseBody;
import org.springframework.web.multipart.MultipartFile;
import com.github.pagehelper.PageInfo;
import com.test.bean.Teacher;
import com.test.service.ITeacherService;
@Controller
@RequestMapping("teacher")
public class TeacherController{
    @Autowired
    private ITeacherService service;
    @RequestMapping("teacherLogin")
    @ResponseBody
    public Map<String,Object>teacherLogin(Teacher teacher, HttpSession session) {
        Boolean falg=false;
        Map<String,Object>map = new HashMap<String,Object>();
        try{
            Teacher teacherMg=service.teacherLogin(teacher);
            if(teacherMg!=null){
                System.out.println(teacherMg);
                if(teacherMg.getStatus()==2){
                    map.put("sAnde","该教师已离职");
                    return map;
                }
                if(teacherMg.getStatus()==3){
                    map.put("sAnde","该教师已被开除");
                    return map;
                }
                if(teacherMg.getIslogin()==2){
                    map.put("sAnde","该账号不能登录");
                    return map;
                }
                session.setAttribute("teacher",teacherMg);
                map.put("sAnde","教师登录成功");
                return map;
            }
        }catch(Exception e){
            e.printStackTrace();
        }
        map.put("sAnde","账号不存在,或密码输入错误");
        return map;
    }
    @RequestMapping("queryAllTeacherMessage")
    @ResponseBody
    public PageInfo<Teacher>queryAllTeacherMessage(Integer pageNum, Integer pageSize){
        PageInfo<Teacher> info = null;
```

第15章 教务管理系统

```java
try {
        info = service.queryAllTeacherMessage(pageNum, pageSize);
    } catch (Exception e) {
e.printStackTrace();
    }
return info;
}
    @RequestMapping("expelTeacher")
    @ResponseBody
public Boolean expelTeacher(Integer tid, Integer status) {
        Boolean flag = false;
try {
//开除老师(不删除仅修改状态)
flag = service.expelTeacher(tid, status);
    } catch (Exception e) {
e.printStackTrace();
    }
return flag;
}
    @RequestMapping("forbidTeacherLogin")
    @ResponseBody
public Boolean forbidTeacherLogin(Integer tid,Integerislogin) {
        Boolean flag = false;
try {
//禁止登录是否可登录 1.是 2否
flag = service.forbidTeacherLogin(tid, islogin);
    } catch (Exception e) {
e.printStackTrace();
    }
return flag;
}
    @RequestMapping("insertTeacher")
public String insertTeacher(Teacher teacher,
    @RequestParam("myFile")MultipartFile myFile,
HttpServletRequest request,Model model) {
try {
//1.设置上传路径
String storPath = request.getServletContext().getRealPath("img");
//2.判断该位置存在不存在
File file = new File(storPath, myFile.getOriginalFilename());
if(!file.getParentFile().exists()) {//如果父路径不存在
file.mkdirs();//则创建目录
}
//3.将文件上传到上传路径
try {
```

```java
                myFile.transferTo(file);
            } catch (IllegalStateException | IOException e) {
                e.printStackTrace();
            }
            teacher.setPhoto(myFile.getOriginalFilename());
            Boolean insertFlag = service.insertTeacher(teacher);
            String msg=insertFlag?"添加成功":"添加失败";
            model.addAttribute("msg", msg);
        } catch (Exception e) {
            e.printStackTrace();
        }
        return "/adminMain.jsp";
    }
    @RequestMapping("queryAllTeacher")
    @ResponseBody
    public Map<String,Object>queryAllTeacher(){
        Map<String,Object> map = new HashMap<String,Object>();
        try{
            List<Teacher> list = service.queryAllTeacher();
            map.put("list", list);
        }catch (Exception e){
            e.printStackTrace();
        }
        return map;
    }
    @RequestMapping("queryTeacherById")
    @ResponseBody
    public Teacher queryTeacherById(HttpSession session){
        Teacher teacher = null;
        try{
            Teacher sessionTeacherMg = (Teacher) session.getAttribute("teacher");
            teacher = service.queryById(sessionTeacherMg.getTid());
        }catch (Exception e){
            e.printStackTrace();
        }
        return teacher;
    }
    @RequestMapping("teacherUpdatePersonalData")
    public String teacherUpdatePersonalData(Teacher teacher,
        @RequestParam("myFile")MultipartFile myFile,
    HttpServletRequest request,Model model,
    HttpSession session){
        try{
            //1.设置上传路径
            String storPath = request.getServletContext().getRealPath("img");
```

```
//2.判断该位置存在不存在
File file = new File(storPath,myFile.getOriginalFilename());
if(!file.getParentFile().exists()){//如果父路径不存在
    file.mkdirs();//则创建目录
}
//3.将文件上传到上传路径
try{
    myFile.transferTo(file);
}catch(IllegalStateException | IOException e){
    e.printStackTrace();
}
teacher.setPhoto(myFile.getOriginalFilename());
Teacher sessionTeacherMg = (Teacher)session.getAttribute("teacher");
teacher.setStatus(sessionTeacherMg.getStatus());
teacher.setIslogin(sessionTeacherMg.getIslogin());
System.out.println("======="+teacher);
Boolean updateFlag = service.teacherUpdatePersonalData(teacher);
String msg=updateFlag?"修改成功":"修改失败";
if(updateFlag){
    Teacher saveSession = service.teacherLogin(teacher);
    session.setAttribute("teacher",saveSession);
}
model.addAttribute("msg",msg);
}catch(Exception e){
    e.printStackTrace();
}
return "/teacherMain.jsp";
    }
}
```

本章小结

本章主要通过一个教务管理系统案例来讲解SSM框架的实际使用。首先对系统的功能、结构等进行了简单的介绍；然后讲解了系统所使用的数据库表；接下来，详细地讲解了系统的环境搭建工作；最后，讲解了功能模块的代码实现。通过本章的学习，读者可以熟练地掌握SSM框架的整合使用，并能熟练地使用SSM框架实现系统功能模块的开发工作。本系统是SSM框架综合使用的案例，读者一定要多加练习，做到熟练编写各个功能模块的实现代码，这样才能将前面所学知识融会贯通。

课后习题

1. 请简述本章教学管理系统的各个层次的组成和作用。
2. 请简述案例中引入 SQL 文件的过程。

参 考 文 献

[1] 黑马程序员. JavaEE 企业级应用开发教程(Spring+Spring MVC+MyBatis)[M]. 北京：人民邮电出版社，2017.

[2] 刘增辉. MyBatis 从入门到精通[M]. 北京：电子工业出版社，2017.

[3] 疯狂软件. Spring+MyBatis 企业应用实战[M]. 北京：清华大学出版社，2017.

[4] 陈恒. JavaEE 框架整合开发入门到实战[M]. 北京：清华大学出版社，2018.

[5] 高洪岩. JavaEE 核心框架实战[M]. 2 版. 北京：人民邮电出版社，2017.

[6] BUILD KUNRNIAWAN, PAUL DECK. Servlet、JSP 和 Spring MVC 初学指南[M]. 林仪明，俞黎敏，译. 北京：人民邮电出版社，2018.

[7] 朱要光. Spring MVC+MyBatis 开发从入门到项目实战[M]. 北京：电子工业出版社，2018.

[8] 郝佳. Spring 源码深度解析[M]. 北京：人民邮电出版社，2015.

[9] 浪潮集团. SSM 框架应用开发与案例实战(Spring+Spring MVC+MyBatis)[M]. 北京：人民邮电出版社，2021.

[10] 李西明. SSM 开发实战教程(Spring+Spring MVC+MyBatis)[M]. 北京：清华大学出版社，2021.

[11] 陈恒，李正光. SSM + Spring Boot + Vue.js 3 全栈开发从入门到实战[M]. 北京：人民邮电出版社，2015.

[12] 千峰教育高教产品研发部. JavaEE(SSM)企业应用实战[M]. 北京：清华大学出版社，2019.